ELECTRONIC COMMUNICATIONS TECHNOLOGY

ELECTRONIC COMMUNICATIONS TECHNOLOGY

Edward A. Wilson

University of Missouri—Rolla

PRENTICE HALL, Englewood Cliffs, New Jersey 07632

Library of Congress Cataloging-in-Publication Data

Wilson, Edward A.
  Electronic communications technology.

  Bibliography: p.
  Includes index.
  1. Telecommunications. I. Title.
TK5101.W47   1989   621.38   88-19715
ISBN 0-13-250333-6

To Christine, Julie, and Heidi

Editorial/production supervision and
  interior design: *Editorial Inc.*
Cover design: *Lundgren Graphics, Ltd.*
Manufacturing buyer: *Robert Anderson*

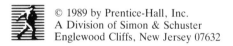
Printed in the United States of America
10 9 8 7 6 5 4 3 2 1

ISBN 0-13-250333-6

Prentice-Hall International (UK) Limited, *London*
Prentice-Hall of Australia Pty. Limited, *Sydney*
Prentice-Hall Canada Inc., *Toronto*
Prentice-Hall Hispanoamericana, S.A., *Mexico*
Prentice-Hall of India Private Limited, *New Delhi*
Prentice-Hall of Japan, Inc., *Tokyo*
Simon & Schuster Asia Pte. Ltd., *Singapore*
Editora Prentice-Hall do Brasil, Ltda., *Rio de Janeiro*

Contents

8   FILTERS AND IMPEDANCE TRANSFORMATION CIRCUITS   122

9   SIGNAL PROCESSING AND CONVERSION CIRCUITS   159

10   MODULATION AND DEMODULATION CIRCUITS   174

11   RF FUNDAMENTALS   185

12   RF COMPONENTS   206

19   SURVEY OF COMMUNICATIONS SYSTEMS   299

Preface

The motivation for development of this text was the perceived need for a pedagogically sound presentation of the field of communications technology, including topics relevant to current and future communications systems. A number of treatises that are available at an advanced engineering level address these topics, but these books require an extensive background in mathematics. A reader with coursework in college-level algebra and trigonometry should have adequate mathematical background for this text; the only other prerequisites are knowledge of AC and DC circuit analysis and some exposure to analog and digital electronics.

I begin by developing a solid understanding of the fundamentals upon which the field of communications is built—spectral analysis, noise theory, sampling techniques, and filtering.

Next comes the topic of modulation. A transition from traditional analog modulation methods to the more current digital techniques is presented with chapters on analog-to-digital and digital-to-analog conversion methods and error control codes. The chapter on digital modulation methods is followed by one on spread spectrum techniques. Although introductory, this material should provide a solid foundation for understanding these systems, which are rapidly gaining importance.

To this point the focus has been on communications system architecture. The text next moves into topics in communications circuit design, including filters, impedance transformation circuits, phase locked loops, companders, A/D and D/A converters, codecs, modulation and demodulation circuits, and RF circuit design.

The text then presents those topics required to complete the reader's understanding of communications systems—RF transmission lines, antennas and propagation.

I conclude with a chapter on fiber optic communications systems and a survey of a wide range of current communications systems. If studied carefully, the material presented here will provide a firm foundation for employment in the communications industry and for the more detailed study of specific systems.

Any work represents the integrated educational background and experience of the author. I wish to thank all who provided the environment and motivation for me to acquire the knowledge necessary for this effort. Special thanks go to four academic experts who kindly reviewed the manuscript of this book: Paul Cary, of the Lincoln Technical Institute; James Williams, of DeVry in Woodbridge; Richard Williams, of the University of Akron; and Robert Greenwood, of the Ryerson Polytechnical Institute.

I owe special thanks to my wife, Christine E. Wilson, who urged me to pursue this project and who assisted me greatly in its completion.

Edward A. Wilson
Rolla, Missouri

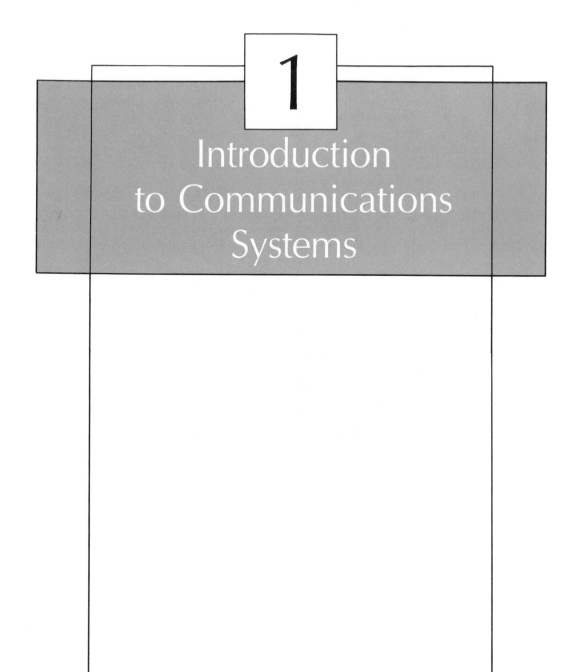

1

Introduction to Communications Systems

One of the most important and certainly one of the most rapidly changing specialties within electronics technology is communications. As recently as the 1970s, a communications specialist had little need for an understanding of digital technology, and microwave techniques were a "black-magic" art practiced only by a select few individuals. Fiber optics was at that time only a laboratory curiosity. Telecommunications was a field dominated by obsolete technology. Packet and spread spectrum techniques were far from becoming practical realities. In the 1970s, satellite communications was in its infancy.

The communications specialist of today needs a clear understanding of the fundamentals upon which the technology is based, including topics such as spectral analysis, noise theory, sampling techniques, and filtering. A strong background in digital technology, including both hardware and software, is also required. Modulation is the process of placing baseband information (voice, data, etc.) onto a so-called "carrier" signal so that it can be transmitted to a receiver that demodulates or extracts the original information; a broad understanding of both analog and digital modulation/demodulation techniques and circuits is a prerequisite for the communications specialist. Since the carrier signal is normally generated at frequencies well above baseband, RF/microwave circuit design, transmission lines, antennas, and propagation are also essential areas of study.

1.1 A GENERALIZED COMMUNICATIONS SYSTEM

The purpose of a communications system is to transfer information between two or more points. In general, a communications system requires a transmitter, a propagation medium, and a receiver. The information to be transferred may be in analog form (e.g., voice or video) or in digital form (computer data, digitized voice, and so on). The transmitter must have some means of superimposing this information onto its output signal (*modulation*). The modulated transmitter signal

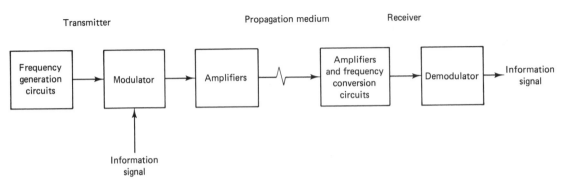

Figure 1-1 Generalized communications system.

must pass through a propagation medium to a receiver. The propagation medium may be free space, an RF transmission line, or a fiber optic cable. The function of the receiver is to select (*filter*) the desired signal, amplify it, and extract the information. A simplified block diagram of a communications system is given in Figure 1-1. The detail within the illustrated blocks is the subject of this text.

1.2 PREREQUISITE BACKGROUND

One of the most fundamental concepts in communications is the relationship between a signal described mathematically (or observed physically) in the time domain and its corresponding characteristics in the frequency domain. A trivial example that may be used to illustrate this correspondence is the sine wave signal first encountered in AC circuit analysis. This signal may be represented mathematically as

$$v(t) = E_m \sin (2\pi f t + \phi) = E_m \sin (\omega t + \phi)$$

In this expression, $v(t)$ is the voltage as a function of time, E_m is the peak voltage, sin is the trigonometric sine function, f is the frequency in Hertz, ω is the radian frequency ($\omega = 2\pi f$), t is time, and ϕ is the phase angle. This mathematical expression describes the signal in the time domain. If one were to plot this expression, the familiar sine curve would result. If the voltage waveform described by this expression were applied to the vertical input of an oscilloscope, a sine wave would be displayed on the CRT screen. Electrical signals (or waveforms) in the time domain may be described mathematically (as a function of time) and observed physically (as a function of time) on an oscilloscope.

The symbol f in the above expression represents the frequency of the sine wave signal. If f is set equal to 1000 KHz and E_m is set to an appropriate level, the signal can be applied to the antenna terminals of an AM broadcast receiver and one will receive the signal when the receiver is tuned to 1000 KHz. The receiver responds to the signal in the frequency domain. A more sophisticated instrument that displays signals in the frequency domain on a CRT screen is the spectrum analyzer. Methods have been developed to write mathematical expressions for signals in the frequency domain. This subject area is known as spectral analysis, and its mathematical methods are called Fourier methods, after the French mathematician who developed them. The Fourier series is used to express periodic time domain functions in the frequency domain, and the Fourier transform is used to express non-periodic time domain functions in the frequency domain.

At this point in your study of communications, it is sufficient to realize that any signal expressed mathematically or observed physically in the time domain may be expressed mathematically or observed physically in the frequency domain. The converse is also true. The oscilloscope may be used to physically observe signals in the time domain and the spectrum analyzer may be used to physically observe signals in the frequency domain.

The limiting factor in the performance of all communications systems is noise. If noise were not present, only an infinitesimal amount of signal power would be required to communicate over an infinitely long communications link. Noise in communications systems is the result of a number of man-made and natural factors. An example of man-made noise is the electromagnetic energy radiated by the arcing commutator brushes on some types of electric motors. Lightning is an often-encountered source of natural noise.

One of the most important sources of noise in communications systems is thermal noise or other atomic noise that can be characterized as thermal noise. These types of noise are the result of electron motion in electronic devices and are always present when such devices are operated above a temperature of absolute zero (0 K or $-273°C$). Low noise performance is a major concern of RF amplifier designers. The ratio of signal power to noise power determines the quality of received signal in an analog communications system and affects the bit error rate (the probability that a given bit of information will be received incorrectly) in a digital communications system.

Although communications signals such as speech and video are inherently analog in nature, digital communications systems generally offer improved performance over analog systems. If an analog signal is to be transmitted over a digital communications system, then it must be digitized by an analog-to-digital (A/D) converter. An A/D converter samples the analog signal at appropriate intervals and forms a digital representation of the amplitude at the sampling instant. A fundamental problem in communications system design is how frequently the analog signal must be sampled in order to obtain an accurate digital representation of the signal; this question is answered by the sampling theorem, which will be discussed in Chapter 2.

Another important concept is that of filtering. A filter may be thought of as a circuit that attenuates signals over an undesired frequency range while passing signals in a desired frequency range with minimal attenuation. Filters are normally designed for specific frequency domain characteristics, but at times their time domain performance is also of concern to communications system designers.

1.3 COMMUNICATIONS SYSTEM DESIGN

Once the fundamental topics of spectral analysis, noise, sampling, and filtering are mastered, a number of additional subjects must be studied before a complete understanding of the field of communications can be developed. These areas may be broadly grouped under the headings of modulation techniques, RF circuit design, transmission lines, antennas, and propagation. These topics are addressed in detail in subsequent chapters.

EXERCISES ━━

1. Generate a list of all of the applications of communications technology of which you are aware.
2. In the laboratory, use a 1000 KHz signal generator, oscilloscope, and AM broadcast receiver to illustrate the time domain vs. frequency domain relationships for this signal.

2

Communications Fundamentals

Each specialized area of electronics technology is founded upon a number of basic concepts. The fundamental concepts of communications include spectral analysis (time domain–frequency domain relationships), noise theory, sampling techniques, and filtering. Spectral analysis, noise theory, and sampling techniques will be treated in this chapter. Filtering will be introduced here but will be covered in Chapter 8. Sophisticated mathematical techniques beyond the scope of this text are required to derive many of the results presented in this chapter, but their use and interpretation do not require this level of mathematics. It is extremely important to develop a physical "feel" for the principles presented in this chapter— the ability to manipulate the associated mathematical expressions is of little value without this feel.

2.1 FOURIER SERIES

A series is often used in mathematics to approximate a complex function by a sum of simple terms. A series approximation normally requires an infinite number of terms to completely eliminate the error between the original function and the approximation. In most commonly used series approximations, however, only a few terms are required for the approximation to converge sufficiently to give a reasonably small error. An example of a series expansion that could be used in a computer program to approximate the trigonometric sine function is

$$\sin x = x - \frac{x^3}{3!} + \frac{x^5}{5!} - \frac{x^7}{7!} + \ldots .$$

where x is expressed in radians. The ! symbol in the above expression indicates the factorial, i.e., $5! = 5 \cdot 4 \cdot 3 \cdot 2 \cdot 1$. This series requires only a few terms for convergence, especially if x is small.

Another important series is the Fourier series, used to approximate a periodic waveform in electronics. A periodic waveform is an electric signal with an amplitude versus time characteristic that repeats in a period T. Examples of periodic waveforms include the rectangular wave, the triangular wave, and of course the sine wave. Figure 2-1 illustrates a periodic rectangular waveform. For this waveform, A is its amplitude, T is its period and τ is the pulse width. This rectangular waveform may be approximated by a Fourier series.

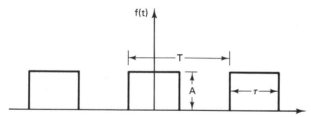

t   **Figure 2-1**   A rectangular waveform.

The general form for the Fourier series is

$$f(t) = \frac{a_0}{2} + \sum_{n=1}^{\infty} \left[a_n \cos \left(\frac{2\pi n t}{T} \right) + b_n \sin \left(\frac{2\pi n t}{T} \right) \right]$$

In this rather complex-looking expression, $\sum_{n=1}^{\infty}$ is simply a mathematical symbol to indicate that an index n is allowed to take on values of 1, 2, 3, 4 . . . up to infinity and that the terms following the symbol are summed. The coefficients a_0, a_n and b_n (known as Fourier coefficients) will be determined by the waveform the Fourier series is to approximate. The symbol T is the period of the waveform and $f(t)$ simply indicates that the Fourier series is a function of time. The expression becomes clear if we write out the first four terms ($n = 1,2,3$) of this series:

$$f(t) = \frac{a_0}{2} + \left[a_1 \cos \left(\frac{2\pi t}{T} \right) + b_1 \sin \left(\frac{2\pi t}{T} \right) \right] + \left[a_2 \cos \left(\frac{4\pi t}{T} \right) \right.$$

$$\left. + b_2 \sin \left(\frac{4\pi t}{T} \right) \right] + \left[a_3 \cos \left(\frac{6\pi t}{T} \right) + b_3 \sin \left(\frac{6\pi t}{T} \right) \right] + \ldots$$

The Fourier series may be written in a slightly different form by noting that

$$\omega_0 = 2\pi f_0 = \frac{2\pi}{T}$$

The symbol f represents frequency in Hz, while ω_0 is known as the radian frequency. The above expression in terms of radian frequency ω_0 is

$$f(t) = \frac{a_0}{2} + [a_1 \cos \omega_0 t + b_1 \sin \omega_0 t] + [a_2 \cos 2 \omega_0 t$$

$$+ b_2 \sin 2 \omega_0 t] + [a_3 \cos 3 \omega_0 t + b_3 \sin 3 \omega_0 t] + \ldots$$

This form illustrates the most fundamental reason for studying the Fourier series. It implies that any periodic waveform may be approximated by a Fourier series that is a summation of sine and cosine terms at the fundamental frequency of the waveform ω_0 and its harmonics $2\omega_0$, $3\omega_0$, etc. Stated differently, it implies that any periodic waveform (such as the rectangular waveform illustrated in Figure 2-1) is simply the sum of sinusoidal waves of various amplitudes and phases (recall that the cosine function is the sine function with a phase shift of 90°). If a periodic signal is applied to the input of a spectrum analyzer (an instrument that displays signals in the frequency domain), the instrument will indicate the presence of power at ω_0, $2\omega_0$, $3\omega_0$ and higher harmonics. Understanding the Fourier series is the first step in developing a solid foundation in understanding the time domain–frequency domain relationships of communications signals.

It was indicated earlier that the Fourier coefficients a_0, a_n, and b_n were dependent upon the waveform to be approximated. The remainder of this section will be devoted to a discussion of the Fourier series for several important waveforms.

The Fourier coefficients for the rectangular waveform of Figure 2-1 are

$$a_0 = \frac{2A\tau}{T}$$

$$a_n = \frac{2A\tau \sin\left(\dfrac{\pi n \tau}{T}\right)}{T\left(\dfrac{\pi n \tau}{T}\right)}$$

$$b_n = 0 \text{ for all terms since the waveform}$$
shown is symmetrical about $t = 0$

The first four terms of the Fourier series for the rectangular waveform of Figure 2-1 are

$$f(t) = \frac{A\tau}{T} + \frac{2A\tau}{T} \frac{\sin\left(\dfrac{\pi\tau}{T}\right)}{\left(\dfrac{\pi\tau}{T}\right)} \cos\left(\frac{2\pi t}{T}\right) + \frac{2A\tau}{T} \frac{\sin\left(\dfrac{2\pi\tau}{T}\right)}{\left(\dfrac{2\pi\tau}{T}\right)} \cos\left(\frac{4\pi t}{T}\right)$$

$$+ \frac{2A\tau}{T} \frac{\sin\left(\dfrac{3\pi\tau}{T}\right)}{\left(\dfrac{3\pi\tau}{T}\right)} \cos\left(\frac{6\pi t}{T}\right) + \ldots$$

Example 2-1

A periodic 1 KHz rectangular waveform (symmetrical about $t = 0$) has a pulse width of 500 μsec and an amplitude of 5 volts. Compute the first four terms of the Fourier series for this waveform.

Solution

The frequency of this waveform is $f = 1 \times 10^3$ Hz, which means that the period T is

$$T = \frac{1}{f} = \frac{1}{1 \times 10^3} = 10^{-3} \text{ sec}$$

The ratio $\dfrac{\tau}{T}$ which appears in the series is

$$\frac{\tau}{T} = \frac{500 \times 10^{-6}}{10^{-3}} = 500 \times 10^{-3} = 0.5$$

Recall that the arguments of the sine and cosine functions in the series are in radians.

The first four terms of the Fourier series for this waveform are

$$f(t) = (5)(0.5) + (2)(5)(0.5) \frac{\sin (0.5\pi)}{0.5\pi} \cos (2\pi \times 10^3 t)$$

$$+ (2)(5)(0.5) \frac{\sin (\pi)}{\pi} \cos (4\pi \times 10^3 t)$$

$$+ (2)(5)(0.5) \frac{\sin (1.5\pi)}{1.5\pi} \cos (6\pi \times 10^3 t)$$

This expression becomes

$$f(t) = 2.5 + 3.18 \cos (2\pi \times 10^3 t) - 1.06 \cos (6\pi \times 10^3 t)$$

Note that the $n = 2$ term is zero and that the $n = 3$ term is negative.

The waveform of Example 2-1 is a special case of the rectangular waveform known as a square wave, since the pulse width is exactly half of the period, or the "on" time equals the "off" time in the waveform. In this example, the frequency spectrum of this waveform consists of a dc (zero frequency) term, a term at 1 KHz and a term at 3 KHz. Additional frequency components are present at odd multiples of 1 KHz. Components at even multiples of 1 KHz are absent, because the waveform is a square wave.

Example 2-2

A periodic 1 KHz rectangular waveform has a pulse width of 100 μsec and an amplitude of 0.5 volts. Calculate the first six terms of the Fourier series for this waveform.

$$T = 10^{-3} \text{ sec}$$

$$\frac{\tau}{T} = \frac{100 \times 10^{-6}}{10^{-3}} = 100 \times 10^{-3} = 0.1$$

$$f(t) = 0.5 + \frac{0.1 \sin (0.31)}{0.31} \cos (2\pi \times 10^3 t) + \frac{0.1 \sin (0.63)}{0.63} \cos (4\pi \times 10^3 t)$$

$$+ \frac{0.1 \sin (0.94)}{0.94} \cos (6\pi \times 10^3 t) + \frac{0.1 \sin (1.24)}{1.24} \cos (8\pi \times 10^3 t)$$

$$+ \frac{0.1 \sin (1.55)}{1.55} \cos (10\pi \times 10^3 t)$$

$$f(t) = 0.5 + 0.98 \cos (2\pi \times 10^3 t) + 0.94 \cos (4\pi \times 10^3 t)$$

$$+ 0.86 \cos (6\pi \times 10^3 t) + 0.76 \cos (8\pi \times 10^3 t)$$

$$+ 0.65 \cos (10\pi \times 10^3 t) + \ldots$$

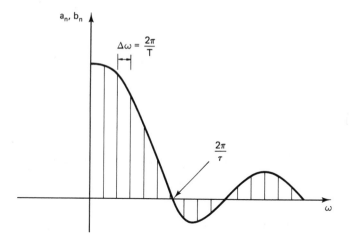

Figure 2-2   Amplitude plot for spectral components of rectangular waveform.

This waveform has a narrower pulse width than that of Example 2-1 and as a result has a spectrum where the amplitude of the fundamental frequency ω_0 and its first few harmonics ($2\omega_0$, $3\omega_0$, $4\omega_0$, $5\omega_0$) decrease slowly.   Note also that the even harmonics are not zero in this case, since the waveform is no longer a square wave.

These two examples illustrate the computational work required to develop the Fourier series for a waveform.   The nature of these computations lends itself to solution on a digital computer.   If a computer is used to calculate a large number of terms for a problem such as Example 2-2 and the amplitudes are plotted, a spectrum similar to Figure 2-2 will be observed.   The "envelope" of these spectral components has the shape of the well-known "sinc" function with $x \geq 0$.   The sinc function is defined as

$$\text{sinc } x = \frac{\sin x}{x}$$

The sinc function is sketched in Figure 2-3.   Referring back to Figure 2-2, note that the first zero crossing occurs at a frequency of

$$\omega = n\omega_0 = \frac{2\pi}{\tau}$$

Note also that most of the signal power lies between $\omega = 0$ and $\omega = 2\pi/\tau$.   These observations lead to a fundamental concept of communications: As pulses in a pulse train (or rectangular waveform) are made shorter, the first zero crossing moves out in frequency, and thus the frequency spread is greater.   Note also that the spacing between spectral components in Figure 2-2 is

$$\Delta\omega = \frac{2\pi}{T}$$

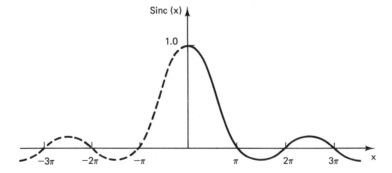

Figure 2-3 Sinc function versus x.

Note that as the period T is made shorter or alternately the frequency is made higher, the frequency lines move out.   This indicates that signal power in the higher frequency ranges increases.

Consider the triangular waveform sketched in Figure 2-4.   The Fourier coefficients for this waveform are

$$a_0 = 0$$

$$a_n = 0 \text{ (all terms)}$$

$$b_n = \frac{-A}{n\pi}$$

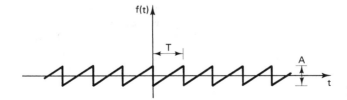

Figure 2-4 Triangular waveform.

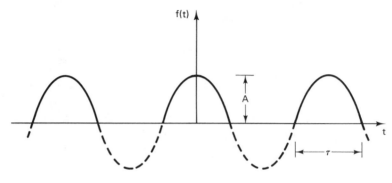

Figure 2-5 Half-cosine waveform.

The Fourier series for this waveform will be developed in a problem at the end of the chapter.

A waveform of great importance to communications as well as all fields of electronics is the *half-cosine* or *rectified cosine* as illustrated in Figure 2-5. If $\tau = T/2$ in this waveform, the output is identical to that of a half-wave rectifier used in power supplies. The Fourier coefficients for this waveform are

$$a_0 = \frac{4A\tau}{\pi T}$$

$$a_n = \frac{A\tau}{T} \frac{\sin\left[\frac{1}{2}\left(\frac{2n\tau}{T} - 1\right)\right]}{\frac{1}{2}\left(\frac{2n\tau}{T} - 1\right)} + \frac{\sin\left[\frac{1}{2}\left(\frac{2n\tau}{T} + 1\right)\right]}{\frac{1}{2}\left(\frac{2n\tau}{T} + 1\right)}$$

$$b_n = 0 \text{ (all terms)}$$

The Fourier series for this waveform will also be investigated in the problems.

Fourier coefficient tables are available in a number of handbooks. Readers familiar with calculus and complex variables may wish to consult a communications text that assumes this level of mathematical background to learn the details of Fourier coefficient computation.

2.2 FOURIER TRANSFORM

Not all waveforms encountered in communications are periodic. A periodic waveform, in fact, conveys no information because it is 100% predictable. The Fourier transform was developed to provide a link between the time domain and the frequency domain for non-periodic waveforms. The Fourier transform is useful in a number of diverse areas besides communications, however. Typical applications include linear system analysis, antennas, optics, statistics, quantum physics, and in the solution of partial differential equations. As with the Fourier series, derivation of the Fourier transform is beyond the scope of this text. Fourier transform results are useful, however, in further developing an understanding of time domain–frequency domain relationships.

Figure 2-6 illustrates a single pulse centered at the origin. The Fourier transform for this pulse is

$$F(\omega) = \frac{A\tau \sin(\omega\tau/2)}{\omega\tau/2}$$

In this expression $F(\omega)$ is the Fourier transform, A is the amplitude of the pulse in the time domain, τ is the pulse width in the time domain, and ω is the radian frequency. The frequency domain behavior of the pulse is illustrated in Figure 2-7. Note that $F(\omega)$ for a single pulse is a continuous function whose shape is the

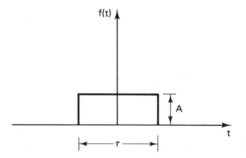

Figure 2-6 Pulse waveform.

sinc function, whereas a Fourier series for a series of pulses is a set of discrete frequency lines within an envelope whose shape is the sinc function. There is a relationship between the Fourier transform and the Fourier series coefficients such that the Fourier transform may be thought of as a set of Fourier series coefficients where the period T of the periodic function is approaching infinity. An infinite period implies a single pulse in the case of a rectangular waveform. Note also that the behavior of $F(\omega)$ is shown for negative frequencies; negative frequencies are not physically realizable and are a by-product of the mathematics used to derive the Fourier transform.

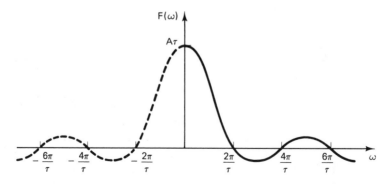

Figure 2-7 Frequency domain behavior of pulse waveform.

Example 2-3

A single pulse waveform is applied to a spectrum analyzer. The analyzer indicates a frequency domain response with the shape of the sinc function, a maximum amplitude equivalent to 10 mv and a first zero crossing of 1 KHz. Calculate the amplitude and pulse width of the applied pulse in the time domain.

Solution

The first zero crossing occurs at

$$\omega = 2\pi f = \frac{2\pi}{\tau}$$

or

$$\tau = \frac{1}{f} = \frac{1}{10^3} = 10^{-3} \text{ sec } = 1 \text{ msec}$$

The maximum amplitude of the Fourier transform for a pulse is

$$F(\omega)_{\text{max}} = A\tau$$

or

$$A = \frac{F(\omega)_{\text{max}}}{\tau} = \frac{10 \times 10^{-3}}{10^{-3}} = 10 \text{ volts}$$

The applied pulse has an amplitude of 10 volts and a duration of 1 msec.

Consider the triangular pulse in the time domain illustrated in Figure 2-8. The Fourier transform for this function is

$$F(\omega) = \frac{A^2 \sin^2 (\omega\tau/4)}{(\omega\tau/4)^2}$$

Figure 2-9 illustrates a waveform that is often called a tone burst. During the time interval τ, the waveform is represented by $A \cos \omega_0 t$ and outside of this interval it is zero.

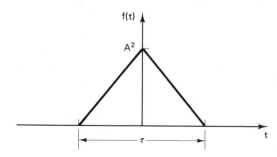

Figure 2-8 Triangular pulse waveform.

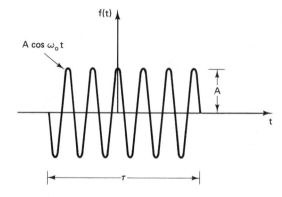

Figure 2-9 Tone burst waveform.

The Fourier transform for the tone burst is

$$F(\omega) = \frac{A^2\tau}{2}\left\{\frac{\sin\left[\tau/2\ (\omega + \omega_0)\right]}{\tau/2\ (\omega + \omega_0)} + \frac{\sin\left[\tau/2\ (\omega - \omega_0)\right]}{\tau/2\ (\omega - \omega_0)}\right\}$$

The behavior of the triangular pulse and the tone burst in the frequency domain will be investigated in the problems at the end of Chapter 2.

Fourier transform tables for the more common time functions are available in a number of handbooks. Calculation methods for the Fourier transform are included in advanced communications textbooks.

2.3 FAST FOURIER TRANSFORM

Many waveforms encountered in industry cannot be satisfactorily defined by mathematical expressions, yet their frequency domain behavior is of great interest. In some cases the need arises to obtain the frequency domain behavior of data that is being collected in the time domain in "real time." The analytical methods discussed thus far are useless for this application.

The discrete Fourier transform was developed for this purpose. In a discrete Fourier transform, the time domain signal is sampled at discrete points in time and the transform is computed from an algorithm coded into a digital computer. Although the discrete Fourier transform has been known for many years, little use was made of it (even with the advent of high-speed digital computers), because its computation time was proportional to N^2, where N is the number of sample points. For any reasonable number of sample points, the computational time was unacceptably long. In 1965, all of this changed when Cooley and Tukey published the now well-known "An Algorithm for the Machine Calculation of Complex Fourier Series" in the *Mathematics of Computation*. This algorithm reduced the computing time to a value proportional to $N\log_2 N$ rather than N^2. The algorithm is commonly called the fast Fourier transform or FFT and is in use worldwide in numerous diverse applications and fields.

Derivations of the discrete Fourier transform and the related fast Fourier transform are well beyond the scope of this text, but as with the Fourier series and the Fourier transform, knowledge of the derivation is not necessary for application of the results. The FFT is available as a subroutine in most scientific subroutine libraries at large computer centers and is available as a part of several statistical packages available for personal computers. The subroutine is compact and requires only about 50 lines of code in a high-level language such as FORTRAN or Pascal. The algorithm has also been imbedded (coded into firmware) in a number of instruments utilized for data acquisition.

For readers with access to an FFT subroutine, an exercise has been included in the problems at the end of Chapter 2.

2.4 SPECTRUM ANALYZERS

An electronic instrument of great value to the communications specialist is the spectrum analyzer, which is used to analyze signals in the frequency domain. It is basically a swept receiver that provides a visual display of amplitude versus frequency. The spectrum analyzer displays the power spectrum of a given waveform. Figure 2-10 is a photograph of the Hewlett-Packard model 8566B 100 Hz to 300 GHz spectrum analyzer and Figure 2-11 is a photograph of the Tektronix model 494AP 10 KHz to 325 GHz portable spectrum analyzer. These two spectrum analyzers are highly complex instruments each costing approximately $50,000.

A spectrum analyzer must be capable of making absolute frequency and amplitude measurements, operate over a large dynamic range in amplitude, have high sensitivity, and provide high resolution of frequency and amplitude. Spectrum analyzers are used to observe the spectral purity of CW signals and to measure frequency conversion products when two signals are mixed, modulation characteristics, pulse spectra, electromagnetic interference (EMI), noise, and the frequency response of amplifiers and/or filters. Frequency response measurements require an external signal generator that tracks the sweep of the spectrum analyzer.

The Fourier analyzer provides more precise and rapid measurements of low-frequency signals than does a standard swept spectrum analyzer. It calculates the frequency domain response using the fast Fourier transform algorithm programmed into its internal microprocessor.

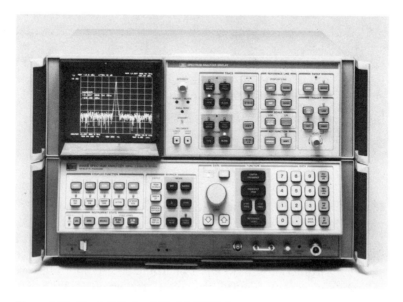

Figure 2-10 Hewlett-Packard Model 8566B Spectrum Analyzer (photo courtesy of Hewlett Packard Company).

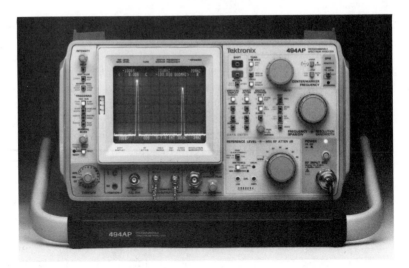

Figure 2-11   Tektronix Model 494AP Portable Spectrum Analyzer (courtesy Tektronix, Inc.).

Spectrum analyzer measurements will be referred to frequently throughout the remainder of this text, as they are essential for communications analysis, design, and troubleshooting.

2.5 NOISE

Noise is the primary limiting factor in the performance of any communication system. The information carrying capacity I of a communications link (sometimes called channel capacity) is measured in bits/sec and is given by the Hartley-Shannon law:

$$I = B \log_2 \left(\frac{P_R}{P_N} + 1 \right) = 1.44 \, B \ln \left(\frac{P_R}{P_N} + 1 \right)$$

In this expression, B is the usable information bandwidth of the link, P_R is the power received and P_N is the noise power. It is apparent from the Hartley-Shannon law that the only ways to increase the information-carrying capacity of a communications link are to increase its bandwidth, to increase the ratio of the received power P_R to the noise power P_N, or to increase both. Observe that for a given information capacity, bandwidth may be "traded off" for the received signal power to noise power ratio.

Example 2-4

A satellite communications link has a bandwidth of 24 MHz. The signal power received is 40 microwatts and the receiver noise power is 2 microwatts. Calculate the information-carrying capacity of this link.

Solution

$$I = 1.44 \ B \ \ln \left(\frac{P_R}{P_N} + 1 \right)$$

$$I = (1.44)(2.4 \times 10^7) \ \ln \ (21)$$

$$I = 1.05 \times 10^8 \ \text{bits/sec} = 105 \ \text{Mbits/sec}$$

The received power in a communications link is a function of the power transmitted, the transmission line loss, antennas, and the propagation path between transmitter and receiver. Noise power is generated both within the communications system and outside it. External sources of noise include the sun, the moon, galactic noise, cosmic noise, sky noise, atmospheric noise and man-made noise. Internal noise arises from intermodulation, shot noise, flicker noise, quantum noise, and thermal noise. Shot noise and flicker noise are the most significant noise sources in vacuum tube circuits, although each type may contribute to the total internal noise generated in semiconductor circuits. In most systems, shot noise and flicker noise may be treated as if they were part of the thermal noise mechanism. Quantum noise is not important for RF communications systems, but is important in optical communications systems. Thermal noise (and shot and flicker noise treated as thermal noise), however, is of great importance to RF communications systems and will be treated in more detail in the remainder of this section.

The science of thermodynamics explains that electrons in a resistive device are always in a state of random motion and that the kinetic energy of these electrons is proportional to T, the absolute temperature. This energy produces voltage fluctuations across the terminals of the device. These fluctuations have a mean value of zero, but a nonzero rms value. This noise source is known as thermal noise or Johnson noise and is often called "white noise," because it contains energy at all frequencies in equal amounts as white light contains all colors in equal amounts.

Thermal noise may be modeled by the equivalent circuit in Figure 2-12. In Figure 2-12, $e_n(t)$ represents the noise voltage, R represents the resistance of the device, and jX represents any reactance that may be present. The open-circuit rms noise voltage is given by

$$\overline{e_n} = \sqrt{4kTBR}$$

In this expression, $\overline{e_n}$ is the open-circuit rms noise voltage in volts, k is Boltzman's

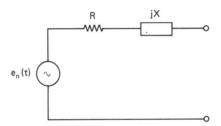

Figure 2-12   Equivalent circuit for thermal noise source.

constant (1.38×10^{-23} watt-sec/K), T is the temperature in Kelvins (K = °C + 273), B is the bandwidth in Hertz, and R is the resistance in ohms.

Example 2-5

A 2.7 KΩ resistor is operated at a temperature of 50°C.   Calculate the rms noise voltage if the measurement bandwidth is 1 MHz.

Solution

$$\overline{e_n} = \sqrt{4kTBR}$$

$$= \sqrt{(4)(1.38 \times 10^{-23})(323)(1 \times 10^6)(2.7 \times 10^3)}$$

$$= 6.94 \times 10^{-6} \text{ volts} = 6.94 \text{ microvolts}$$

Electric circuit theory indicates that maximum power will be transferred from a two-port network if it is conjugate matched.   In the equivalent circuit of Figure 2-12, a conjugate match allows calculation of the maximum noise power $P_{n\ max}$. Figure 2-13 illustrates the conjugate match situation.   In this circuit, the reactances cancel, and the maximum noise power $P_{n\ max}$ delivered to the load resistor R is

$$P_{n\ max} = \frac{\overline{e_n}^2}{4R} = kTB$$

In this expression, $P_{n\ max}$ is in watts.

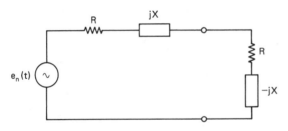

Figure 2-13   Conjugate match to thermal noise equivalent circuit.

Example 2-6

Calculate the maximum noise power produced by a 2.7 KΩ resistor operated at a temperature of 50°C.   The measurement bandwidth is 1 MHz.

Solution

$$P_{n\ max} = kTB$$

$$= (1.38 \times 10^{-23})(323)(1 \times 10^{6})$$

$$= 4.46 \times 10^{-15} \text{ watts}$$

Note that the value of the resistor does not enter into the noise power computation.

To account for other sources of noise such as shot noise, flicker noise or quantum noise, the concept of effective noise temperature, T_e, is introduced. T_e is defined from the expression for maximum noise power:

$$P_{n\ max} = kT_eB$$

or

$$T_e = \frac{P_{n\ max}}{kB}$$

Devices such as low-noise RF amplifiers are sometimes rated by their effective noise temperature T_e.

An alternate means of describing the noise performance of RF amplifiers is by the noise factor F.   Noise factor is defined as

$$F = \frac{S_i/N_i}{S_o/N_o}\bigg|_{T=T_0}$$

In this expression, S_i represents input signal power to the amplifier, S_o represents output signal power, N_i represents input noise power, and N_o is the output noise power.   The ratios S_i/N_i and S_o/N_o are called signal-to-noise ratios and are of great importance in communications.   Note that noise factor is defined at the standard temperature of $T_0 = 290$ K.   Noise factor may be thought of as the degradation in signal-to-noise ratio caused by the amplifier.

Example 2-7

The signal input to an RF amplifier is 40 microwatts and the noise power at the amplifier input is 2 microwatts.   The signal-to-noise ratio at the output is measured to be 14.   Assume $T = 290$ K.   Compute the noise factor of this amplifier.

Solution

$$\frac{S_i}{N_i} = \frac{40 \times 10^{-6}}{2 \times 10^{-6}} = 20$$

$$F = \frac{S_i/N_i}{S_o/N_o} = \frac{20}{14} = 1.43$$

An alternate definition of noise factor may be developed as follows:

$$\frac{S_o}{S_i} = G \text{ (gain)}$$

$$N_i = kTB$$

$$F = \frac{S_i/N_i}{S_o/N_o} = \frac{S_i N_o}{S_o N_i} = \frac{N_o}{kTBG}\bigg|_{T=T_0}$$

Example 2-8

A microwave amplifier has a gain of 20 dB, a bandwidth of 30 MHz, and a noise factor of 2.5. Calculate the noise power at the output of the amplifier if it is operated at 30°C.

Solution

$$F = \frac{N_o}{kTBG}$$

$$N_o = FkTBG$$

$$\text{dB} = 10 \log \frac{P_o}{P_i} = 10 \log G$$

$$G = \log^{-1} \frac{\text{dB}}{10}$$

$$G = \log^{-1} 2 = 100$$

$$N_o = (2.5)(1.38 \times 10^{-23})(303)(3 \times 10^7)(100)$$

$$N_o = 3.14 \times 10^{-11} \text{ watts}$$

Noise factor is normally expressed in dB and is called noise figure, *NF*:

$$NF = 10 \log F$$

The noise factor of a low-noise RF amplifier or entire receiver may be converted to a *receiver-effective noise temperature* by the following expression:

$$T_e = (F - 1) \, T_0$$

In this expression, T_e is the receiver-effective noise temperature, F is the noise factor, and T_0 is 290 K (13°C)—a standard temperature.   In this expression, if $F = 1$, which means the receiver (or amplifier) is noiseless, $T_e = 0$ K.   Noise temperature is often used instead of noise factor or noise figure to describe the performance of low-noise amplifiers.

Example 2-9

An 8–18 GHz microwave amplifier's specification sheet indicates that its noise temperature is 150 K when it is operated at 25°C.   Calculate the amplifier's noise figure.

Solution

$$T_e = (F - 1) \, T_0 \qquad (T_0 = 290 \text{ K})$$

$$F = \frac{T_e}{T_0} + 1$$

$$F = \frac{150}{290} + 1 = 1.52$$

$$NF = 10 \log F$$

$$NF = 10 \log (1.52)$$

$$NF = 1.82 \text{ dB}$$

An additional convenience is that noise temperatures may be added directly. For example, if T_i is the noise temperature of the input termination to a receiver with noise temperature T_e, the total noise temperature is

$$T_t = T_i + T_e$$

In practical communications systems, noisy subsystems are often cascaded and the overall noise figure or noise temperature of the cascaded system is of interest.   As an example, in a receiver, a transmission line is often cascaded with one or two RF amplifiers, followed by a mixer, which is followed by one or more IF amplifiers.

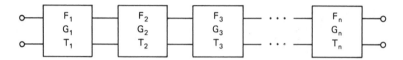

Figure 2-14   A system of n noisy subsystems in cascade.

Figure 2-14 illustrates the general situation.   For a cascaded system with n stages, the overall noise factor is

$$F_{in} = F_1 + \frac{F_2 - 1}{G_1} + \frac{F_3 - 1}{G_1 G_2} + \cdots \cdot \frac{F_n - 1}{G_1 G_2 \ldots G_{n-1}}$$

Expressed in terms of noise temperature, the overall temperature is

$$T_{in} = T_1 + \frac{T_2}{G_1} + \frac{T_3}{G_1 G_2} + \cdots \cdot \frac{T_n}{G_1 G_2 \ldots G_{n-1}}$$

Note that the major contributor to noise factor or noise temperature in a cascaded system is the first stage.

Example 2-10

A 2–4 GHz amplifier with a noise figure of 2 dB and a gain of 20 dB is cascaded with another 2–4 GHz amplifier with a noise figure of 6 dB and a gain of 30 dB. Calculate the overall gain and noise figure for the cascaded system.

Solution

$$NF = 10 \log F$$

$$F = \log^{-1} \frac{NF}{10}$$

$$F_1 = \log^{-1} (0.2) = 1.58$$

$$F_2 = \log^{-1} (0.6) = 3.98$$

$$G_{dB} = 10 \log G$$

$$G = \log^{-1} \left(\frac{G_{dB}}{10} \right)$$

$$G_1 = \log^{-1} (2) = 100$$

$$G_2 = \log^{-1} (3) = 1000$$

$$F_{12} = F_1 + \frac{F_2 - 1}{G_1}$$

$$F_{12} = 1.58 + \frac{3.98 - 1}{100} = 1.58 + 0.03 = 1.61$$

$$NF_{12} = 10 \log F_{12} = 10 \log (1.61) = 2.07 \text{ dB}$$

$$G_{12} = G_1 G_2 = (100)(1000) = 10^5 = 50 \text{ dB}$$

or

$$G_{12\text{dB}} = G_{1\text{dB}} + G_{2\text{dB}} = 20 \text{ dB} + 30 \text{ dB} = 50 \text{ dB}$$

Note that the rather noisy second stage has little effect upon the overall noise figure because of the high-gain G_1 of the relatively low-noise preamplifier.

Example 2-11

Two identical 4–8 GHz amplifiers with a gain of 20 dB each are cascaded and the overall effective noise temperature of the two-amplifier system is measured to be 150 K. Calculate the noise figure of an individual amplifier.

Solution

$$T_{in} = T_1 + \frac{T_2}{G_1} \qquad G_1 = 20 \text{ dB} = 100 \qquad T_1 = T_2 = T$$

$$150 = T + \frac{T}{100}$$

$$150 = 1.01 \, T$$

$$T = 148.5 \text{ K} = T_1 = T_2$$

$$T_1 = (F_1 - 1) \, T_0$$

$$F_1 = \frac{T_1}{T_0} + 1$$

$$F_1 = \frac{148.5}{290} + 1$$

$$F_1 = 1.51$$

$$NF_1 = 10 \log F_1$$

$$NF_1 = 10 \log (1.51)$$

$$NF_1 = 1.79 \text{ dB}$$

As previously indicated, in a receiver the first noisy subsystem is the transmission line from the antenna to the input of the receiver. The transmission line is a lossy matched attenuator with a physical temperature T_p and a loss factor L.

The loss factor is the inverse of gain and is defined as

$$L_t = \frac{P_i}{P_o}$$

Note that the loss factor will always be greater than 1.0 for an actual transmission line. The noise factor for a lossy transmission line is

$$F_t = 1 + (L_t - 1)\frac{T_p}{T_0}$$

where T_0 is the reference temperature, 290 K. The effective noise temperature at the input of a lossy transmission line is then

$$T_t = (L_t - 1)T_p$$

The expressions previously given may be used for a transmission line in a cascaded system by making the following substitution:

$$G_1 = \frac{1}{L_t}$$

Example 2-12

A 460 MHz communications receiver has a noise figure of 2 dB. A 100-meter length of coaxial cable transmission line connects the receiver to an antenna. The loss of the transmission line at 460 MHz is measured at 6 dB. The temperature of the transmission line is 20°C. Calculate the overall noise figure of this communications system.

Solution

$$L_t = 6 \text{ dB} = 3.98$$

$$G_t = \frac{1}{L_t} = 0.25$$

$$F_t = 1 + (L_t - 1)\frac{T_p}{T_0}$$

$$F_t = 1 + (3.98 - 1)\frac{293}{290} = 4.01$$

$$NF_r = 2 \text{ dB} \qquad F_r = 1.58$$

$$F_{12} = F_1 + \frac{F_2 - 1}{G_1} = F_t + \frac{F_r - 1}{G_t}$$

$$F_{12} = 4.01 + \frac{1.58 - 1}{0.25} = 4.01 + 2.32$$

$$F_{12} = 6.33$$

$$NF_{12} = 8.01 \text{ dB}$$

Note that when a receiver with an excellent noise figure of 2 dB is preceded with a lossy transmission line, the overall system noise figure is seriously degraded.

Antenna noise temperature is another important factor in communications system design. Antenna noise temperature is an equivalent temperature defined as follows:

$$T_a = \frac{P_a}{k\Delta f}$$

In this expression, P_a is the noise power intercepted by the antenna over the frequency range of interest Δf. Antenna noise temperature has no direct relationship to the physical temperature of the antenna. An antenna's physical temperature might be only 200 K ($-73°C$), yet its noise temperature may be several thousand Kelvins if it is pointed toward the sun, or as low as 30 K if pointed towards a "quiet" region of space.

Antenna noise temperature T_a, transmission line equivalent noise temperature T_t and receiver equivalent noise temperature T_r may be combined to calculate a system noise temperature T_s. It is common practice to specify the system noise temperature at the input terminals of the receiver. At the receiver input terminals, the antenna temperature T_a will be multiplied by the gain of the transmission line:

$$T_a' = \frac{T_a}{L}$$

At the receiver input terminals, the noise temperature at the input of the transmission line will also be multiplied by the gain of the transmission line:

$$T_t' = \frac{T_t}{L_t} = \frac{(L_t - 1)\, T_p}{L_t} = \left(1 - \frac{1}{L_t}\right) T_p$$

The receiver equivalent noise temperature at the input to the receiver is just T_r.

The overall expression for system noise temperature, T_s, referred to the receiver input terminals is

$$T_s = \frac{T_a}{L_t} + \left(1 - \frac{1}{L_t}\right) T_p + T_r$$

If it is desirable, the system noise temperature may be specified at the input of the transmission line rather than at the input of the receiver. The above expression for T_s must then be multiplied by L:

$$T_s' = T_a + (L_t - 1)\, T_p + L T_r$$

Example 2-13

A 4 GHz microwave receiving system has an antenna temperature of 250 K, a waveguide transmission line with a loss of 4 dB, and a receiver noise temperature of 150 K.   The entire system is operated at a physical temperature of 20°C.   Calculate the overall system noise figure at the receiver input terminals.

Solution

$$L_t = 4 \text{ dB} = 2.51$$

$$T_s = \frac{T_a}{L_t} + \left(1 - \frac{1}{L_t}\right) T_p + T_r$$

$$T_s = \frac{250}{2.51} + \left(1 - \frac{1}{2.51}\right) 293 + 150$$

$$T_s = 99.6 + 176.3 + 150$$

$$T_s = 425.9$$

$$T_s = (F_s - 1) T_0$$

$$F_s = \frac{T_s}{T_0} + 1$$

$$F_s = \frac{425.9}{290} + 1 = 2.47$$

$$NF_s = 10 \log F_s$$

$$NF_s = 10 \log (2.47)$$

$$NF_s = 3.9 \text{ dB}$$

Note that the receiver noise temperature without the antenna and transmission line was 150 K or a noise figure of 1.82 dB.

2.6 NOISE MEASUREMENTS

The most useful definition of noise factor for purposes of measurement is

$$F = \left. \frac{N_o}{kTBG} \right|_{T=T_0}$$

Consider a device with an impedance-matched input termination, Z_s, at a temperature of T_s.   The device is noisy and contributes excess noise power N_a to the output.   This situation is illustrated in Figure 2-15.   The noise at the output of the device is:

$$N_o = N_a + kT_s BG$$

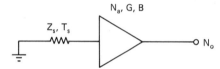

Figure 2-15 Noisy device with input termination.

A plot of this equation is given in Figure 2-16. In this plot, two temperatures T_{s_1} and T_{s_2} with corresponding noise output powers N_1 and N_2 are indicated. The plot is observed to be a straight line with slope kBG and y-intercept N_a. Since a straight line is determined by two points, the illustrated line may be found by measuring the output noise power at two input termination temperatures, T_{s_1} and T_{s_2}. The equation of the straight line is the expression for noise output power, N_0, which may be used in the expression for noise factor F. Note again the noise factor is defined at the standard temperature, $T_0 = 290$ K.

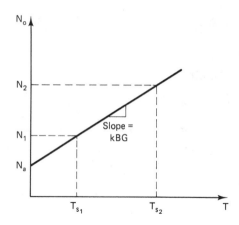

Figure 2-16 Plot of expression for noise output power.

Example 2-14

A 1.3 GHz, L-band microwave amplifier is terminated with its input impedance of 50 ohms. The input termination is cooled to 250 K and the output noise power is measured to be 2.3 nanowatts. The input termination is then heated to 300 K and the output noise power is measured to be 2.5 nanowatts. Calculate the noise figure of this amplifier.

Solution

The slope of the line is:

$$kBG = \frac{\Delta N_o}{\Delta T_s} = \frac{2.5 \times 10^{-9} - 2.3 \times 10^{-9}}{300 - 250} = \frac{2.0 \times 10^{-10}}{50} = 4.0 \times 10^{-12}$$

N_a may be found by substitution of T_{s_1} and N_1 or T_{s_2} and N_2 into the equation for N_o. Using T_{s_1} and N_1:

$$N_o = N_a + kT_s BG$$

or

$$N_a = N_o - kT_s BG$$
$$= 2.3 \times 10^{-9} - (4 \times 10^{-12})(250)$$
$$= 1.0 \times 10^{-9}$$

The equation for the output noise is then

$$N_o = 1.0 \times 10^{-9} + 4 \times 10^{-12}\, T$$

The noise factor is

$$F = \frac{N_o}{kTBG} = \left. \frac{1.0 \times 10^{-9} + 4 \times 10^{-12}\, T}{4 \times 10^{-12}\, T} \right|_{T = T_0}$$

$$F = \frac{1.0 \times 10^{-9} + 1.16 \times 10^{-9}}{1.16 \times 10^{-9}} = 1.86$$

$$NF = 10 \log F = 2.70 \text{ dB}$$

Since it is inconvenient to heat and cool an input termination, noise figure measurements are normally taken with the "cold" temperature, T_{s_1}, equal to the ambient temperature and the "hot" temperature T_{s_2} provided by a noise source such as a diode. Noise sources are characterized by an excess noise ratio ENR which is given by:

$$\text{ENR} = 10 \log \frac{T_{s_2} - T_0}{T_0}$$

The impedances of noise sources are made equal to common input impedances of amplifiers, for example, 50 ohms. When the noise source is turned off, it appears as a matched input termination at ambient temperature, T_{s_1}, and when it is turned on, it appears as a matched input termination of temperature T_{s_2}, which can be determined from the excess noise ratio of the source. Noise figure may be calculated as illustrated in Example 2-14.

Noise figure measurements involve a number of calculations, so the computational power of the microprocessor has been incorporated into instruments known as automatic noise figure meters. These instruments are used in conjunction with an external noise source to provide instantaneous noise figure measurements. Noise figure meters are more accurate than manual methods, because they com-

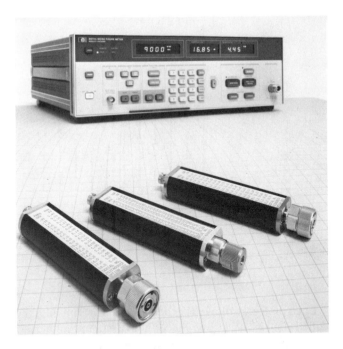

Figure 2-17 Hewlett-Packard Model HP8970A Noise Figure Meter and associated noise sources (photo courtesy Hewlett-Packard Company).

pensate for any noise contribution of the measuring equipment through calibration and can also compensate for any variation in ENR of the noise source as a function of frequency.

Figure 2-17 illustrates the Hewlett-Packard model HP8970A noise figure meter and associated noise sources. The HP8970A can measure noise figure directly from 10 MHz to 1.5 GHz and is usable up to 26.5 GHz with an external mixer. A commonly used noise source, the HP346A, can supply an ENR of approximately 5 dB over a frequency range of 10 MHz to 18 GHz. Other noise source models provide higher ENR and/or frequency coverage. The cost of this system is approximately $15,000.

2.7 SAMPLING THEOREM

The world is primarily analog in nature. Communications signals such as speech and video are examples of analog information that is transmitted over a communications link. Although analog communications systems are available, digital communications systems are often more effective. Analog signals are converted to a digital format by a device known as an analog-to-digital converter (A/D). At the receiving end of the communications link, digital data may be converted back to an analog format by use of a digital-to-analog converter (D/A). An analog signal is converted to a digital format by a process known as *sampling*. Sampling is illustrated in Figure 2-18. The sampled signal consists of voltage levels defined

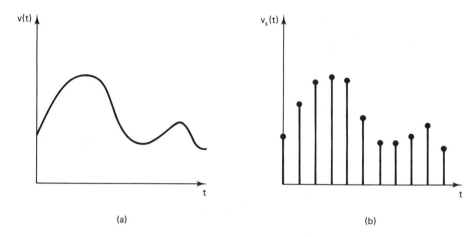

Figure 2-18   Analog signal (a) and sampled signal (b).

at discrete intervals of time. Each discrete voltage is called a sample. The number of samples per unit time is called the *sampling frequency* f_s. The sampling period is $T_s = 1/f_s$. From Figure 2-18b, it is apparent that as the sampling frequency f_s approaches infinity, the sampled signal approaches the original analog signal. On the other hand, as the sampling frequency is decreased toward zero, the sampled signal will diverge from an accurate representation of the original analog signal.

    The sampling theorem states that the minimum sampling frequency, f_s, must be equal to twice the information bandwidth ($BW_I = f_{\max} - f_{\min}$) of the original analog signal to ensure an accurate representation. This minimum sampling rate is called the Nyquist rate after H. Nyquist of Bell Laboratories. The sampling theorem in equation form is

$$f_s \geq 2\ BW_I$$

For baseband signals $f_{\min}$ is usually approximated by zero so that $BW_I = f_{\max}$.

    The expression "accurate representation" implies that the original signal can be recovered from the sampled version by suitable filtering. The highest component, $f_{\max}$, in a signal may be determined analytically by Fourier analysis or experimentally with a spectrum analyzer. Analog signals are often filtered prior to sampling to ensure that no frequency component exceeds the desired $f_{\max}$. In addition, analog signals are often sampled at a frequency somewhat higher than the Nyquist rate to allow a *guard band*. Sampling is often illustrated schematically as shown in Figure 2-19.

    Derivation of the sampling theorem is beyond the scope of this text, but the theorem result will be utilized often in this text and in communications system design.

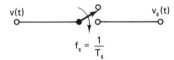

Figure 2-19 Schematic illustration of sampling.

2.8 FILTERING

Filtering, the selection of particular components over a specified range in the frequency domain, is a fundamental concept of communications. The most common filter types are low pass, high pass, band pass, and band reject. Low pass, high pass, and band pass filters have a well-defined *pass band* over which signals applied to the filter experience minimal insertion loss. The insertion loss increases rapidly outside the pass band for these filter types. The band reject filter has a well-defined *stop band* over which signals applied to the filter experience maximum insertion loss. The insertion loss decreases rapidly outside the stop band for this filter type.

A low pass filter response is illustrated in Figure 2-20. The pass band for this filter extends from $\omega = 0$ to $\omega = \omega_c$. The frequency ω_c is called the cutoff frequency and is normally defined as the point where the response curve is 3 dB below the point of minimum insertion loss.

A high pass filter response is illustrated in Figure 2-21. The pass band for this filter extends from $\omega = \omega_c$ to $\omega = \infty$. The frequency ω_c is called the cutoff frequency and likewise is normally defined as the point where the response curve is 3 dB below the point of minimum insertion loss.

A band pass filter response is illustrated in Figure 2-22. The pass band for this filter extends from $\omega = \omega_1$ to $\omega = \omega_2$. ω_1 is the lower 3 dB cutoff frequency and ω_2 is the upper 3 dB cutoff frequency.

A band reject filter response is illustrated in Figure 2-23. The stop band for this filter extends from $\omega = \omega_1$ to $\omega = \omega_2$. The frequencies ω_1 and ω_2 are the points where the response curve is 3 dB above the point of maximum insertion loss.

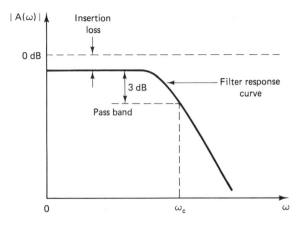

Figure 2-20 Low pass filter response.

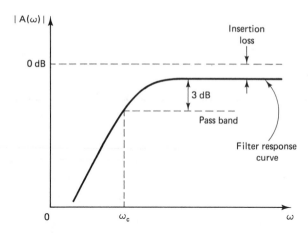

Figure 2-21 High pass filter response.

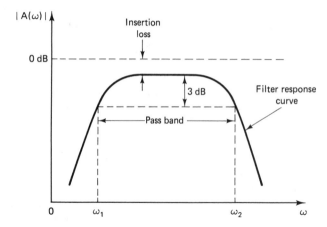

Figure 2-22 Band pass filter response.

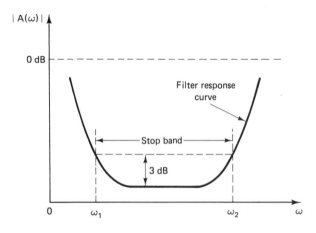

Figure 2-23 Band reject filter response.

The shape of the filter's response in the pass band is also of interest. Certain filter design techniques result in "ripple" in the pass band (or stop band) in exchange for a faster "rolloff" outside of the pass band. Filters will be covered in more detail in Chapter 8.

EXERCISES ━━

1. Calculate the percent error of the series approximation to the sine function given in Section 2.1 for five terms and an angle of 50°.
2. Compute the first five terms of the Fourier series for a periodic 100 KHz triangular waveform with a peak amplitude of 10 volts and where a positive peak occurs at $t = 0$.
3. Compute the first four terms of the Fourier series for a periodic half-cosine waveform with a period of 10 μsec, A = 1.0, and with $T = \tau$ and where a positive peak occurs at $t = 0$.
4. Use a personal computer or programmable calculator to compute points for the Fourier series in Exercise 2 or Exercise 3. Compare your results to the original waveform with a plot.
5. The spectrum of a single triangular pulse has a first zero crossing point of 1 MHz when observed on a spectrum analyzer. Compute the duration of the pulse.
6. A 5-volt, 0.1 second tone burst of frequency 2 KHz is applied to a spectrum analyzer. Write an expression for the frequency domain response observed on the instrument.
7. Use an FFT subroutine to compute the frequency domain response of the tone burst in Exercise 6. Plot your result.
8. A 4.2 GHz satellite downlink signal is applied to a low-noise amplifier with a noise figure of 2 dB. If the signal-to-noise ratio at the amplifier input is 20 dB, compute the signal-to-noise ratio at the amplifier output. Assume the system is operated at 290 K.
9. Compute the effective noise temperature corresponding to a noise figure of 1 dB. Assume a 290 K operating temperature.
10. A 12 GHz amplifier with a noise figure of 1 dB and a gain of 25 dB is cascaded with a second 12 GHz amplifier with a noise figure of 8 dB and a gain of 30 dB. Calculate the overall gain and noise figure for this cascaded system.
11. A 1.3 GHz low-noise amplifier is connected to an antenna with a length of coaxial cable having a loss of 3 dB. The amplifier has a noise figure of 1 dB and the system is operating at 10°C. Compute the overall noise figure of the amplifier and transmission line.
12. A 10 GHz amplifier is terminated with its input impedance of 50 ohms. The input termination is cooled to $-30°C$ and the output noise power is measured to be 2.0×10^{-9} watts. The termination is then heated to $+30°C$ and the output noise power increases by 0.2 nanowatts. Compute the noise figure of this amplifier.

Analog Modulation Methods

Fundamental to communications is the transfer of information between two or more points via a medium of propagation such as free space, wires, coaxial cables, or fiber optic cables.   The information-bearing signal in a communications system is called the baseband signal.   Modulation is the process of superimposing a baseband signal onto a higher frequency carrier signal so that the information may be transmitted efficiently.   Recovery of the baseband signal at the receiving end of the communications link is known as demodulation or detection.   Baseband signals may be transmitted directly over wire links, but the loss involved may be high. Also, it is often desirable to transmit several baseband signals over the same pair of wires—a situation that requires a process known as multiplexing, in which a single carrier is modulated by a number of baseband signals.   A signal may be multiplexed in the time domain (time division multiplexing) or in the frequency domain (frequency division multiplexing).   In theory, baseband signals can be transmitted through free space, but since antennas must be of the order of magnitude of a wavelength, it is not practical even to consider such direct transmission. A carrier may be represented in the time domain as:

$$v_c(t) = A \cos(\omega_c t + \phi)$$

In this expression $v_c(t)$ is the carrier voltage as a function of time, A is the carrier amplitude, ω_c is the radian frequency of the carrier ($\omega_c = 2\pi f_c$), t is time, and ϕ is the phase of the carrier.   This carrier may be modulated by varying A, ω_c, or ϕ.   Modulation methods that vary A are called amplitude modulation, methods that directly vary ω_c are called frequency modulation, and methods that directly vary ϕ are called phase modulation.   If the modulating signal is a continuous function of time, the modulation method is called analog; if the modulating signal has a finite number of possible levels or states, the method is called digital.   Pulse modulation techniques such as pulse amplitude modulation, pulse width modulation and pulse position modulation are analog methods, because the modulating signal varies continuously, while techniques such as frequency shift keying are digital. Digital modulation will be discussed in Chapter 6.

3.1 FREQUENCY TRANSLATION

Frequency translation, sometimes called mixing or heterodyning, is a process often encountered in communications systems.   The process of modulation also creates a frequency translation, so it will be introduced in this section, but the concept of frequency translation is so pervasive throughout communications that it will appear often in this text.

Consider a baseband or modulating signal represented in the time domain as follows ($\phi = 0$):

$$v_m(t) = A_m \cos \omega_m t$$

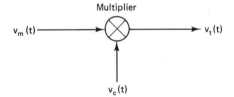

Multiplier

$v_m(t)$ ───────▶ $\otimes$ ───────▶ $v_t(t)$

$v_c(t)$

Figure 3-1 Multiplication of two signals.

A carrier signal may be represented in the time domain as ($\phi = 0$):

$$v_c(t) = A_c \cos \omega_c t$$

It is assumed that $\omega_m \ll \omega_c$, which is the normal case.

Frequency translation may be achieved by multiplying these two signals as illustrated in Figure 3-1:

$$v_t(t) = v_m(t) \cdot v_c(t) = A_m A_c \cos \omega_m t \cos \omega_c t$$

The translation becomes evident if the following trigonometric identity is utilized:

$$\cos \alpha \cos \beta = \frac{1}{2} \cos (\alpha + \beta) + \frac{1}{2} \cos (\alpha - \beta)$$

Using this identity, the translated signal may be written as

$$v_t(t) = \frac{A_m A_c}{2} \cos (\omega_c + \omega_m)t + \cos (\omega_c - \omega_m)t$$

Figure 3-2 illustrates the frequency domain representation of this signal. The translated signal is seen to consist of two signals, one at a frequency of $f_c - f_m$, and the second at a frequency of $f_c + f_m$. Note that this result is for a "perfect" multiplier. A realistic multiplier circuit implemented with electronic components would allow the two input signals at frequencies f_c and f_m also to appear in the output. The baseband signal at f_m may be filtered-out, however, and the carrier at f_c may be suppressed, so the ideal multiplier is a useful concept. The signal illustrated in Figure 3-1 has one component at ($f_c - f_m$), which is called the lower sideband, and the other component at ($f_c + f_m$), called the upper sideband. Since

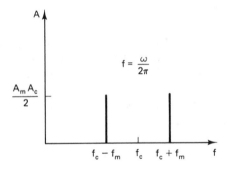

A

$f = \dfrac{\omega}{2\pi}$

$\dfrac{A_m A_c}{2}$

$f_c - f_m \quad f_c \quad f_c + f_m \qquad f$

Figure 3-2 Frequency domain representation of translated signal.

the carrier at f_c is suppressed in an ideal multiplier, the translated signal is called a double sideband suppressed carrier signal (DSBSC or DSB).

Example 3-1

A 15 KHz audio tone with a peak amplitude of 10 volts and a 12 MHz RF signal with a peak amplitude of 20 volts are applied to an ideal multiplier circuit. Determine the nature of the signal(s) present at the output of this device.

Solution

An ideal multiplier would have two sinusoidal signals at its output, one at $(f_c - f_m)$ and one at $(f_c + f_m)$. In this example:

$$f_m = 15 \times 10^3 \text{ Hz}$$

$$f_c = 12 \times 10^6 \text{ Hz}$$

$$f_c - f_m = 11.985 \times 10^6 = 11.985 \text{ MHz}$$

$$f_c + f_m = 12.015 \times 10^6 = 12.015 \text{ MHz}$$

If the baseband signal is a non-periodic signal $m(t)$, band-limited in the frequency domain to a maximum frequency of f_m, multiplication with a carrier will translate the entire baseband signal to an upper sideband component and a lower sideband component centered at f_c. Figure 3-3a illustrates the baseband signal in the frequency domain; Figure 3-3b illustrates the spectrum of the translated signal.

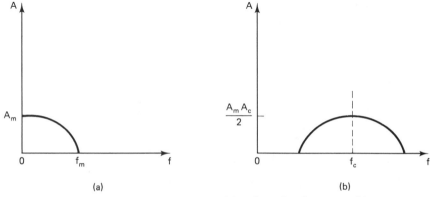

Figure 3-3 Baseband spectrum (a) and translated spectrum (b).

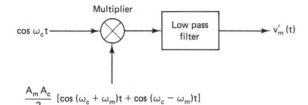

$$\frac{A_m A_c}{2} \, [\cos (\omega_c + \omega_m)t + \cos (\omega_c - \omega_m)t]$$

Figure 3-4   Recovery of baseband signal.

Multiplication by $\cos \omega_c t$ may be used to recover the baseband information from a translated signal. Figure 3-4 illustrates the procedure. The translated signal is

$$v_t(t) = \frac{A_m A_c}{2} \, [\cos (\omega_c + \omega_m)t + \cos (\omega_c - \omega_m)t] = A_m \cos \omega_m t \, A_m \cos \omega_c t$$

Multiplying by $\cos \omega_c t$:

$$\cos \omega_c t \, [A_m \cos \omega_m t \cdot A_m \cos \omega_c t] = A_m \cos \omega_m t \cdot A_c \cos^2 \omega_c t$$

Recall the trigonometric identity

$$\cos^2 \alpha = \frac{1}{2} + \frac{1}{2} \cos 2\alpha$$

Using this identity, the "recovered" signal is

$$\frac{A_m}{2} \cos \omega_m t + \frac{A_c}{2} \cos 2 \, \omega_c t$$

The term $2 \omega_c$ is filtered out by the low-pass filter, leaving only the baseband signal:

$$v_m'(t) = \frac{A_m}{2} \cos \omega_m t$$

This recovery method requires a synchronous signal, $\cos \omega_c t$, at the recovery point. Any phase or frequency shifts would modify the amplitude of the baseband signal. A synchronous signal can sometimes be derived from the received signal with additional circuitry at the receiver.

3.2 AMPLITUDE MODULATION

An amplitude modulated (AM) signal may be generated by multiplying the baseband signal, $m(t)$, with the carrier signal and adding the carrier signal by itself to the product. Figure 3-5 illustrates the procedure. The amplitude modulated signal may be described mathematically as (for $\phi = 0$)

$$v(t) = A_c \cos \omega_c t + m(t) \cos \omega_c t = A_c \, [1 + m(t)] \cos \omega_c t$$

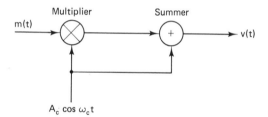

Figure 3-5 Amplitude modulation procedure.

Figure 3-6a illustrates the carrier in the time domain; Figure 3-6b illustrates an arbitrary baseband or modulating signal $m(t)$ in the time domain.   Figure 3-7 illustrates the result of amplitude modulation with these two signals.   The depth of amplitude modulation is described by the percent modulation.   0% modulation indicates an unmodulated carrier, while 100% modulation is normally the maximum modulation possible without overmodulation distortion.   Percent modulation is defined as follows:

$$\% \text{ modulation} = \frac{A_{c(\max)} - A_{c(\min)}}{2\,A_c} \times 100\%$$

In this expression, $A_{c(\max)}$ is the maximum upper envelope amplitude, $A_{c(\min)}$ is the minimum upper envelope amplitude, and A_c is the unmodulated carrier amplitude.

Example 3-2

A baseband signal with a peak positive excursion of 600 mV and a peak negative excursion of -400 mV is used to amplitude modulate a 30 MHz RF carrier with a peak amplitude of 10 volts.   Calculate the percent modulation.

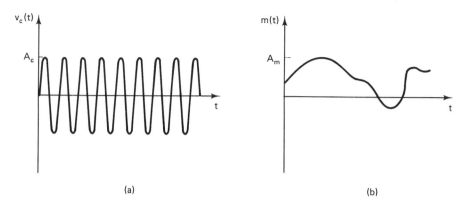

Figure 3-6   Carrier in time domain (a) and baseband signal in time domain (b).

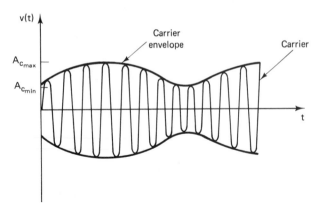

Figure 3-7 Amplitude modulation.

Solution

$$v(t) = A_c [1 + m(t)] \cos \omega_c t$$

$$A_{c(max)} = A_c [1 + m(t)_{max+}] = 10 [1.0 + 0.6] = 16 \text{ volts}$$

$$A_{c(min)} = A_c [1 + m(t)_{max-}] = 10 [1.0 - 0.4] = 6 \text{ volts}$$

$$A_c = 10 \text{ volts}$$

$$\% \text{ modulation} = \frac{A_{c(max)} - A_{c(min)}}{2 A_c} \times 100\%$$

$$= \frac{16 - 6}{20} = \frac{10}{20} = 50\%$$

An amplitude modulated signal's baseband may be recovered or demodulated by a synchronous demodulator as discussed in Section 3.1. A major advantage of amplitude modulation, however, is the ease of recovery of the baseband using only a non-linear device and a low-pass filter. A simple rectifying diode followed by a low-pass filter is a common method of AM demodulation or detection. Demodulation may also be achieved with other non-linear elements such as a square-law device. A square-law device biased at some DC level, A_o, may be described by

$$v_o = k v_i^2$$

If it is assumed that the input to this device is an amplitude modulated signal:

$$v_i = A_o + A_c [1 + m(t)] \cos \omega_c t$$

where A_o is the DC bias. The output of the device would then be

$$v_o = k\{A_o + A_c [1 + m(t)] \cos \omega_c t\}^2$$

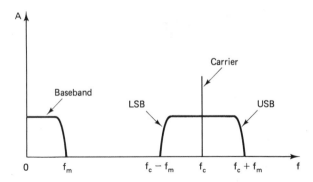

Figure 3-8 Baseband signal and corresponding AM signal.

If the above expression is expanded and it is assumed that all frequencies above baseband are filtered out and the DC term is blocked, the output signal is

$$v_o = kA_c^2 \left[m(t) + \frac{1}{2} m^2(t) \right]$$

The $1/2m^2(t)$ term will introduce distortion, but its effect will be minimal if $1/2m^2(t) \ll |m(t)|$. It is important to note that demodulation may occur unintentionally anywhere in a system when an amplitude modulated signal is passed through a supposedly linear device such as an amplifier that exhibits some degree of non-linear distortion.

The spectrum of a baseband signal and the corresponding amplitude modulated signal is illustrated in Figure 3-8. Note that this spectrum is similar to the DSBSC spectrum in Section 3.1, except that a carrier component is present. Note from Figure 3-8 that the bandwidth of an AM signal is $2f_m$; also note that the upper sideband and the lower sideband carry identical information, so one of the sidebands is redundant. Although it allows the use of simple modulation and demodulation circuitry, AM is very wasteful of both the spectrum and transmitter power and finds little application in modern communications system design. An AM receiver is also quite susceptible to man-made and natural sources of noise since it, by definition, must respond to amplitude variations. The primary importance of AM at present is in the broadcast industry because of the need for compatibility with receivers of older design.

3.3 SINGLE SIDEBAND MODULATION

It was noted in Section 3.2 that redundant information is produced in an amplitude modulated or double sideband signal. The carrier contains no information, and the information contained in the upper sideband is identical to the information contained in the lower sideband. Since an AM or DSB signal occupies excessive spectrum, it is important to consider single sideband modulation or SSB. In SSB

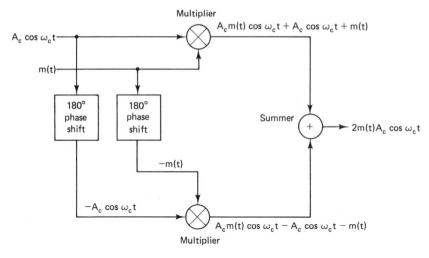

Multiplier

$A_c m(t) \cos \omega_c t + A_c \cos \omega_c t + m(t)$

$A_c \cos \omega_c t$

$m(t)$

| 180° phase shift | 180° phase shift |

Summer

$2m(t)A_c \cos \omega_c t$

$-m(t)$

$-A_c \cos \omega_c t$

$A_c m(t) \cos \omega_c t - A_c \cos \omega_c t - m(t)$

Multiplier

Figure 3-9 Balanced modulator.

either the upper or lower sideband is transmitted, but not both. Single sideband modulation occupies only one-half the spectrum of an AM or DSB signal.

An essential element in single sideband generation is the balanced modulator. Figure 3-9 illustrates the operation of a balanced modulator. Note that the output of a real (as opposed to ideal) multiplier contains the two input signals as well as the desired product signal. The output of the summing junction and thus of the balanced modulator is a true DSB signal as discussed in Section 3-1.

A single sideband signal may be produced from a double sideband signal by filtering out the unwanted sideband. This method of single sideband generation is known appropriately as the filter method and is the most common method of generation. A single sideband signal developed by the filter method is normally generated at a low RF frequency above baseband, since the required filter must have a very sharp cutoff. Sharp cutoff filters are much easier to design at lower RF frequencies. Figure 3-10 illustrates SSB generation.

A single sideband signal may be described mathematically as follows:

$$v(t) = m(t) \cos \omega_c t - m'(t) \sin \omega_c t \text{ (USB)}$$

$$v(t) = m(t) \cos \omega_c t + m'(t) \sin \omega_c t \text{ (LSB)}$$

In these expressions, $m'(t)$ is the baseband or modulating signal with all frequency components shifted in phase by 90°. The above expressions give rise to an alternate

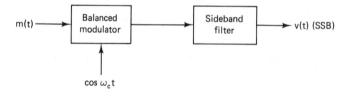

$m(t)$

| Balanced modulator | → | Sideband filter | → $v(t)$ (SSB) |

$\cos \omega_c t$

Figure 3-10 Single sideband generation.

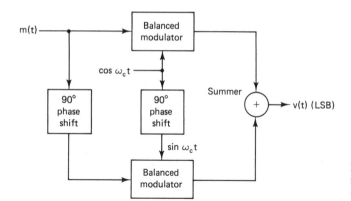

Figure 3-11 Phase-shift method of generating SSB (LSB) signal.

method of generating a single sideband signal known as the phasing or phase-shift method. Figure 3-11 illustrates the phase shift method of generating a lower sideband signal. To generate the upper sideband signal, the output of the bottom balanced modulator would have to be multiplied by -1 (or shifted in phase 180°) prior to summing.

The phase-shift method of generating an SSB signal is not as popular as the filter method, because it is extremely difficult to design a network to shift all components of the baseband signal by 90° while maintaining constant amplitudes. The phase-shift method is also more sensitive to other imperfections introduced by "real" electronic circuitry, such as errors in the carrier phase shift and imperfectly matched balanced modulators.

Synchronous demodulation is necessary to recover the exact baseband information from a single sideband signal. Synchronous detection or demodulation requires a signal at the receiver synchronized with the carrier $\cos \omega_c t$ in both frequency and phase. To assist the receiver in providing this signal, some SSB transmission systems insert a "pilot" carrier slightly above the highest baseband frequency. The receiver circuitry "locks" onto this signal to produce a synchronous signal (injected carrier) for demodulation. Figure 3-12 illustrates synchronous demodulation for a SSB signal. The output of a real multiplier for an LSB signal is the product terms plus the input signals:

$$m(t) \cos^2 \omega_c t + m'(t) \sin \omega_c t \cos \omega_c t + m(t) \cos \omega_c t + m'(t) \sin \omega_c t + \cos \omega_c t$$

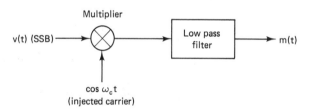

Figure 3-12 Synchronous demodulation of SSB signal.

The first term may be expanded using the trigonometric identity for $\cos^2 \alpha$:

$$m(t) \cos^2 \omega_c t = m(t) \left[\frac{1}{2} + \frac{1}{2} \cos 2\omega_c t \right]$$

Except for the $m(t)/2$ term, all of the other terms at the multiplier output are at frequency ω_c or higher, so the baseband signal may be recovered with a low-pass filter at the multiplier output.

For communications-quality speech transmission, synchronous detection of the SSB signal is not necessary. In this case, the circuit of Figure 3-12 is still used, but the oscillator signal does not have to be exactly at f_c. As long as the oscillator frequency is within a few tens of Hertz of f_c, an understandable speech signal will result at the output of the demodulator circuit. The demodulator circuit is then referred to as a product detector rather than a synchronous detector. It should be noted that synchronous detection is required to demodulate a DSB signal.

3.4 AMPLITUDE COMPANDERED SINGLE SIDEBAND MODULATION

Single sideband modulation has found application in communications systems since the late 1940s. Demodulation of an SSB signal requires an "injection" oscillator at or very close to the frequency of the suppressed carrier. At high carrier frequencies some manual means of "tuning" this inserted carrier is often necessary. This control, which is called a "clarifier" by some manufacturers, presents little difficulty to operators of fixed receivers. Until recently, operators of land mobile communications systems found that this necessary adjustment was unacceptable for their mode of operation. Thus, SSB systems were rarely found in mobile radio systems, despite the considerable spectrum savings of SSB over the FM modulation systems that have been predominantly used for mobile communications.

In 1985, the Federal Communications Commission (FCC) approved an enhancement of SSB called amplitude compandered single sideband (ACSB®) for the land mobile service. ACSB systems have overcome the need for a "clarifier," offer up to a 4:1 improvement in spectrum utilization over FM systems, and feature a 12–15 dB improvement in signal-to-noise ratio over FM.

ACSB systems utilize a device known as a compander. The term compander is derived from the two words compressor and expander. Companders are most often found in digital communications systems, where an analog voice signal is digitized for transmission over the system. The compressor compresses the amplitude of the audio signal prior to transmission, and the expander expands a previously compressed signal at the receiver. The "transfer function" between output and input of the compressor or expander section of a compander is known as the "compander law." Several compander laws are in use. Compander operation will be discussed in more detail in Chapter 4, and compander circuits will be covered in Chapter 9. In ACSB, the compander utilized is typically a 2:1 amplitude compander. A 2:1 amplitude compander's compression section, for example, provides an output peak signal of 1 volt for a 2 volt peak signal input. A 2:1 amplitude expander provides the complementary function, that is, it provides an output of 2 volts peak for a 1 volt peak signal input. Companders are available from a number of manufacturers in integrated circuit form.

ACSB systems provide a 3.1 KHz pilot tone used for carrier synchronization at the receiver and to send signaling and control information. The pilot tone is well filtered and does not interfere with the audio baseband information. The pilot is typically transmitted at a power level of 10 dB below the maximum RF output on voice peaks. In a pure SSB system with no pilot tone, the RF output with no baseband signal would be zero, but with the pilot at 10 dB below audio peaks, the system would have an output of 10% of maximum power output with no baseband signal.

ACSB systems provide a number of advantages over FM systems for land mobile applications, but both of these analog techniques face increased competition from digital radio systems, as will be discussed in Chapter 6.

3.5 FREQUENCY AND PHASE MODULATION

Frequency and phase modulation are two special cases of the more general form of modulation called angle modulation. A carrier signal may in general be represented in the time domain as

$$v_c(t) = \cos(\omega_c t + \phi) = \cos \theta(t)$$

The argument of the cosine function $\theta(t)$ is an angular function of time. Angle modulation varies either the frequency or phase of $\theta(t)$ linearly with the modulating signal. In phase modulation, the phase of the carrier ϕ is made to vary linearly with the modulating signal. Variations in ϕ directly vary the phase of $\theta(t)$. In frequency modulation, the instantaneous frequency is made to vary linearly with the modulating signal. The instantaneous frequency is the rate of change of $\theta(t)$ with respect to time or its derivative with respect to time. Frequency and phase modulation are so closely related that examination of an FM or PM signal gives no information as to which type of modulation is in use. The modulation type can, however, be determined from an examination of the transmitter circuitry. The remainder of this section will emphasize FM, but the results can also be made to apply to PM.

FM is an extremely popular method of analog modulation primarily because of its noise immunity. The average power of an FM carrier does not vary with modulation and is the same as the average power of the unmodulated carrier. This constant carrier amplitude characteristic of FM allows the use of limiter circuits at the receiver that effectively remove all amplitude variations in the received signal. This ability to limit the received signal contributes to the superior noise performance of FM, since most noise on the received signal arises in the form of amplitude variations. Figure 3-13 illustrates an FM signal in the time domain.

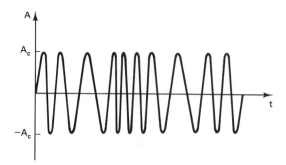

Figure 3-13   FM signal in time domain.

An FM signal is best analyzed by considering a sinusoidal modulating signal. Actual modulating signals are more complex, of course, and would have to be handled by using Fourier transform methods.   Fortunately, the results obtained by assuming sinusoidal modulation are generally applicable to more complex modulating waveforms.   The frequency modulated carrier resulting from a $\sin \omega_m t$ modulating signal may be expressed as follows:

$$v(t) = \cos (\omega_c t + \beta \sin \omega_m t)$$

In this expression, ω_m is the radian frequency of the modulating sinusoid; β is known as the modulation index, defined as

$$\beta = \frac{\Delta \omega}{\omega_m}$$

In this expression, $\Delta \omega$ is known as the peak frequency deviation. $\Delta \omega$ is proportional to the amplitude of the modulating signal and depends upon modulator circuit constants.

A special case of FM, important in applications such as land mobile communications, is narrow-band FM.   Narrow-band FM occurs under the following condition:

$$\beta \ll \frac{\pi}{2}$$

Normally the condition $\beta < 0.2$ is chosen as the defining criterion.

Narrow-band FM may be analyzed by utilizing the following trigonometric identity:

$$\cos (A + B) = \cos A \cos B - \sin A \sin B$$

Setting $A = \omega_c t$ and $B = \beta \sin \omega_m t$, the expression for the FM signal becomes

$$v(t) = \cos \omega_c t \cos (\beta \sin \omega_m t) - \sin \omega_c t \sin (\beta \sin \omega_m t)$$

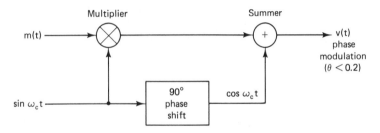

Figure 3-14   Generation of a phase modulated signal.

This expression may be simplified by observing the following characteristics of the sine and cosine functions:

$$\cos \alpha \approx 1 \text{ and } \sin \alpha \approx \alpha \text{ if } \alpha \ll \frac{\pi}{2}$$

The FM signal for $\beta \ll \pi/2$ becomes approximately

$$v(t) = \cos \omega_c t - \beta \sin \omega_m t \sin \omega_c t$$

$$= \cos \omega_c t - \frac{\beta}{2} [\cos (\omega_c - \omega_m)t - \cos(\omega_c + \omega_m)t]$$

Note that this signal is similar to but not identical with the product modulator output discussed previously. It is similar to an AM signal in that it has two sidebands displaced $\pm \omega_m$ from a carrier at ω_c, and the bandwidth of the signal is $2f_m$. The narrow-band FM signal differs from the AM signal, however, in that its amplitude is constant, and the two sideband terms are in phase opposition (180° phase shift) with each other. In the narrow-band FM signal, the phase and thus the instantaneous frequency varies.

A frequency or phase modulated signal may, in fact, be generated using a multiplier as illustrated in Figures 3-14 and 3-15. The frequency modulation scheme illustrated in Figure 3-15 differs from the phase modulation scheme only by the addition of the integrator block in the FM case. An integrator is a simple circuit that behaves somewhat like a low-pass filter. In this application the baseband

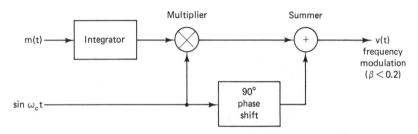

Figure 3-15   Generation of a frequency modulated signal.

signal must first be integrated to ensure that the instantaneous modulator output frequency varies linearly with the modulating signal.

Example 3-3

A 150 MHz carrier is FM modulated by a sinusoidal signal of 10 KHz. The frequency deviation is 1 KHz. Calculate the modulation index, the bandwidth and the sideband frequencies of this signal.

Solution

$$\beta = \frac{\Delta\omega}{\omega_m} = \frac{\Delta f}{f_m} = \frac{1 \times 10^3}{10 \times 10^3} = 0.1$$

$$BW = 2f_m = 20 \text{ KHz}$$

Sidebands at $f_c \pm f_m$ = 150.01 MHz and 149.99 MHz

Wide-band FM finds application primarily in the broadcast industry and is defined as follows:

$$\beta > \frac{\pi}{2}$$

In the case of wide-band FM, the previously utilized approximations are invalid, and a more general approach to the analysis of the FM signal must be taken. The original FM signal expression was:

$$v(t) = \cos \omega_c t \cos (\beta \sin \omega_m t) - \sin \omega_c t \sin (\beta \sin \omega_m t)$$

The above expression may be expanded into a rather involved Fourier series that requires the use of a set of mathematical functions known as Bessel functions. The expression becomes

$$v(t) = J_0(\beta) \cos \omega_c t - J_1(\beta) \cos (\omega_c - \omega_m)t - \cos (\omega_c + \omega_m)t$$

$$+ J_2(\beta) \cos (\omega_c - 2\omega_m)t - \cos (\omega_c + 2\omega_m)t$$

$$- J_3(\beta) \cos (\omega_c - 3\omega_m)t - \cos (\omega_c + 3\omega_m)t$$

$$+ \dots$$

In the above expression, $J_n(\beta)$ represents values of the Bessel functions of the first kind. Table 3-1 gives selected values of the Bessel functions, while Figure 3-16 illustrates the behavior of $J_0(\beta)$.

The wide-band FM signal is seen to consist of a carrier term plus an infinite number of sidebands. Examination of Table 3-1 indicates, however, that only the first few sidebands are significant even for large β. From Table 3-1, it also can be observed that if β is less than 0.2, the only significant sidebands are the first

TABLE 3-1 SELECTED VALUES OF BESSEL FUNCTIONS OF THE FIRST KIND

β	$J_0(\beta)$	$J_1(\beta)$	$J_2(\beta)$	$J_3(\beta)$	$J_4(\beta)$	$J_5(\beta)$	$J_6(\beta)$
0.0	1.000	0.000	0.000	0.000	0.000	0.000	0.000
0.1	0.998	0.050	0.001	0.000	0.000	0.000	0.000
0.2	0.990	0.010	0.005	0.000	0.000	0.000	0.000
0.4	0.960	0.196	0.020	0.001	0.000	0.000	0.000
1.0	0.765	0.440	0.115	0.020	0.002	0.000	0.000
1.4	0.567	0.542	0.207	0.050	0.009	0.001	0.000
2.0	0.224	0.577	0.353	0.129	0.034	0.007	0.001
2.4	0.003	0.520	0.431	0.198	0.064	0.016	0.003
3.0	−0.260	0.339	0.486	0.309	0.132	0.043	0.011
3.4	−0.364	0.179	0.470	0.373	0.189	0.072	0.022
4.0	−0.397	−0.066	0.364	0.430	0.281	0.132	0.049
4.4	−0.342	−0.203	0.250	0.430	0.336	0.182	0.076
5.0	−0.178	−0.328	0.047	0.365	0.391	0.261	0.131

set, which confirms the approximate analysis given for narrow-band FM. As β is increased from the narrow-band case, the number of significant sidebands increases and the amplitude of the carrier decreases. At β ≈ 2.4, the carrier amplitude goes to zero. As β is increased beyond 2.4 the amplitude of the carrier will increase again, but its phase will be shifted by 180°. A useful approximation for the bandwidth required by an FM signal is:

$$BW = 2(\beta + 1)f_m = 2(\Delta f + f_m)$$

In this expression, Δf is the frequency deviation and f_m is the modulating frequency. Although this result was derived by assuming a sinusoidal modulating signal, it is useful for estimating bandwidth if f_m is the lowest modulating frequency in a baseband signal. The number of significant sideband pairs is β + 1. It should be noted that, when β << 1, the wide-band FM equation for $v(t)$ reduces to the approximate equation developed for $v(t)$ in the narrow-band FM case.

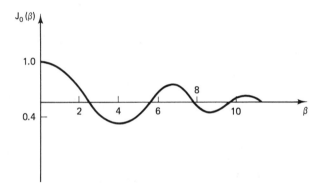

Figure 3-16 Behavior of the Bessel function $J_0(\beta)$.

Example 3-4

In the U.S.A., the FCC allows a frequency deviation of 75 KHz for commercial FM broadcasting in the 88 MHz to 108 MHz band. If a 15 KHz tone is transmitted over such a system, calculate the relative amplitude of the carrier and all significant sidebands.

Solution

$$\beta = \frac{\Delta f}{f_m} = \frac{75 \times 10^3}{15 \times 10^3} = 5$$

Carrier $\alpha\ J_0(\beta) = J_0\ (5) = -0.178$

1st pair $\alpha\ J_1(\beta) = J_1\ (5) = -0.328$

2nd pair $\alpha\ J_2(\beta) = J_2\ (5) = 0.047$

3rd pair $\alpha\ J_3(\beta) = J_3\ (5) = 0.365$

4th pair $\alpha\ J_4(\beta) = J_4\ (5) = 0.391$

5th pair $\alpha\ J_5(\beta) = J_5\ (5) = 0.261$

6th pair $\alpha\ J_6(\beta) = J_6\ (5) = 0.131$

Example 3-5

A commercial FM broadcast station transmits audio signals from 100 Hz to 15 KHz. Calculate the bandwidth of the signal and the number of significant sidebands.

Solution

$$\beta = \frac{\Delta f}{f_m}$$

$$\beta_{low} = \frac{75 \times 10^3}{10^2} = 750$$

$$\beta_{high} = \frac{75 \times 10^3}{15 \times 10^3} = 5$$

For $\beta = 750$:

$$BW = 2(\beta + 1)f_m = (2)(751)(100) = 150.2\ \text{KHz}$$

Number of significant sideband pairs $= \beta + 1 = 751$

For $\beta = 5$:

$$BW = 2(\beta + 1)f_m = (2)(6)(15 \times 10^3) = 180.0\ \text{KHz}$$

Number of significant sideband pairs $= 6$

Note that there are fewer sideband pairs for higher frequency modulating signals, but they are spaced farther apart than those produced by a low-frequency modulating signal which produces many more pairs.

A frequency modulated signal is most often generated by a device known as a voltage-controlled oscillator (VCO). In this device, the modulating signal $m(t)$ is applied to a circuit element that changes its capacitance with voltage. This VVC or voltage variable capacitor is utilized in the frequency-determining part of the VCO's circuit. Often it is desirable to increase the frequency deviation Δf available from a VCO, which may be accomplished using a device known as a frequency multiplier. The frequency multiplier increases the carrier frequency and the deviation by a factor n. Frequency multiplier circuits are covered in Chapter 13.

An FM signal may be demodulated by a discriminator circuit, a ratio detector circuit, or (most often in modern systems) a phase locked loop circuit. A discriminator is basically a frequency-selective network that converts frequency variations into amplitude variations, and a ratio detector is a modified discriminator that does not respond to rapid amplitude variations.

A phase locked loop is a circuit that locks onto and tracks the carrier and provides an output signal proportional to the frequency deviation that is the demodulated baseband signal. Discriminators and ratio detectors will be discussed in Chapter 10, and phase locked loops will be discussed in Chapters 9 and 10.

3.6 FREQUENCY DIVISION MULTIPLEXING

Often in communications systems many baseband signals must be transmitted over a single communications channel. Multiplexing is the term given to this technique. The multiplexing technique utilized in continuous-time analog communications systems is frequency division multiplexing (FDM). Time division multiplexing is primarily utilized in sampled data analog and digital communications systems, and will be introduced in Section 3.7.

Figure 3-17 illustrates a technique for generating a frequency division multiplexed signal. In Figure 3-17, $m_1(t), m_2(t), \ldots m_n(t)$ represent the band-limited baseband signals, and $\cos \omega_1 t, \cos \omega_2 t, \ldots \cos \omega_n t$ represent carriers. The carrier frequencies $\omega_1, \omega_2, \ldots \omega_n$ are chosen to prevent overlap with modulation and also to allow for a "guard band" between each of the signals. Figure 3-18 illustrates the frequency domain representation of the signal at the output of the summer in Figure 3-17. Figure 3-17 shows DSB modulation being used on each subcarrier. SSB is also employed in FDM systems to conserve spectrum. The carriers $f_1, f_2, \ldots f_n$ are normally chosen to be much lower in frequency than the ultimate signal to be transmitted over the communications channel, so a frequency translator or up-converter is required to translate the multiplexed signal up to a range suitable for transmission. It should be noted that some wire and cable communications channels do not require the use of an up-converter.

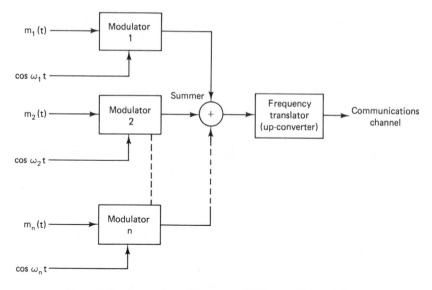

Figure 3-17 Generation of frequency division multiplexed signal.

Figure 3-19 illustrates a technique for separating the composite frequency division multiplexed signal into individual baseband signals at the receiving end. After down-conversion, each of the bandpass filters in Figure 3-19 selects a carrier and the modulation about that carrier. After separation, each signal is demodulated and filtered at baseband to remove the unwanted demodulator products and noise.

Example 3-6

Three baseband signals, each band limited to 4 KHz, are to be transmitted over a single 4 GHz communications link. Devise a scheme to generate and demodulate a frequency division multiplexed signal for this application.

Solution

Arbitrarily choose $f_1 = 90$ KHz, $f_2 = 100$ KHz and $f_3 = 110$ KHz, which allows for a 2 KHz guard band. Note that the composite signal extends from 86 KHz to 114 KHz. The composite signal is up-converted to 4 GHz, transmitted over the communications link and down-converted to 86 KHz to 114 KHz. Bandpass filters

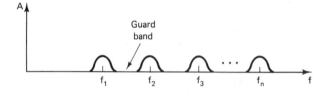

Figure 3-18 Spectrum of frequency division multiplexed signal.

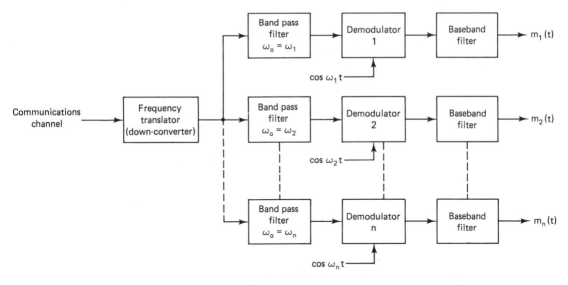

Figure 3-19 Frequency division multiplex demodulator.

at 90 KHz, 100 KHz, and 110 KHz, each with a bandwidth of 8 KHz, separate the signals, which are applied to individual demodulators.   The locally generated carrier inputs to the demodulators are 90 KHz, 100 KHz, and 110 KHz respectively. Each demodulator is followed by a baseband filter with a bandwidth of 4 KHz.

Frequency division multiplexing is used extensively in telecommunications and satellite communications systems, but is now being replaced by time division multiplexing as these systems are converted to digital.   The vast majority of frequency division multiplex systems utilize FM modulation of the main carrier by a multiplexed signal consisting of a number of SSB signals at different subcarrier frequencies.

3.7 TIME DIVISION MULTIPLEXING

Section 3.6 illustrated a method for transmitting many baseband signals over a single communications channel by utilizing frequency translation techniques.   We now give a brief introduction to another multiplexing technique, time division multiplexing (TDM).

Time division multiplexing is based upon the sampling theorem, discussed in Section 2.7.   The sampling theorem states that a signal band-limited to f_m must be sampled at a frequency of at least $2f_m$ if distortion is to be avoided.   In a time division multiplexed system, a number of signals are sampled in sequence and interleaved in time to form the TDM signal.   Figure 3-20 illustrates a possible time

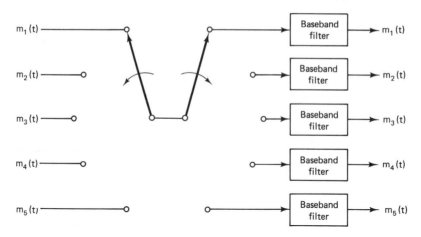

Figure 3-20   Time division multiplex system.

division multiplexing system.   The two commutator switches illustrated in Figure 3-20 must be synchronized, and each baseband signal must be sampled at the Nyquist rate or greater.   In actual systems, electronic switches are utilized rather than mechanical switches due to reliability, size, power, and speed considerations.

A time division multiplexed signal may be transmitted directly over a wire communications channel or may be translated to a higher frequency for transmission.   The bandwidth required for a TDM signal is

$$BW = Nf_m$$

In this expression, N is the number of baseband signals and f_m is the maximum baseband frequency.   It is interesting to note that this bandwidth is identical to that required to frequency division multiplex N baseband signals using single sideband modulation of each subcarrier.

Time division multiplexing is normally used in digital communications systems but also finds application in pulse modulation systems that are analog in nature, such as pulse amplitude modulation (PAM) systems, pulse width modulation (PWM) systems, and pulse position modulation (PPM) systems.

3.8 PULSE AMPLITUDE MODULATION

Pulse amplitude modulation (PAM) is a form of modulation based upon the sampling theorem.   In PAM, an analog signal is sampled at the Nyquist rate or greater and a pulse is generated with its amplitude proportional to the amplitude of the analog signal.   Figure 3-21 illustrates the concept of pulse amplitude modulation. The baseband signal is sampled at a rate $R_s \geq 2f_m$, where f_m is the maximum baseband frequency.   The output of the sampler is a train of impulses with period $T_s = 1/R_s$.

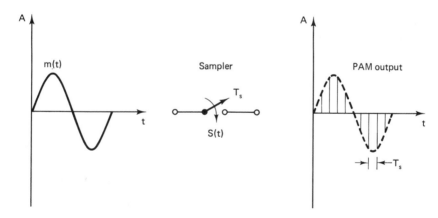

Figure 3-21   Pulse Amplitude Modulation system.

A more realistic method of generating a PAM signal is illustrated in Figure 3-22.   In Figure 3-22, $m(t)$ is the modulating signal; $S(t)$ is the sampling function, illustrated in Figure 3-23.

Figure 3-24 illustrates a typical PAM output from a multiplier circuit as in Figure 3-21.   Note that the pulse tops have been "flattened" at the output of the multiplier by the use of sample and hold circuitry.

Pulse amplitude modulation is most often utilized in conjunction with a time division multiplexing system as discussed in Section 3.7.   The baseband signal may be recovered from a pulse amplitude modulation signal by passing the PAM signal through a holding circuit and then through a low-pass filter with bandwidth equal

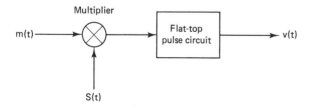

Figure 3-22   Generation of realistic PAM signal.

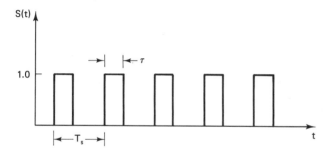

Figure 3-23   The sampling function.

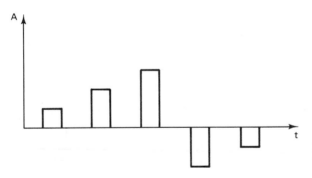

Figure 3-24 Pulse Amplitude Modulation output.

to f_m, the highest baseband frequency.   PAM is considered an analog modulation method because the pulses may assume any amplitude.   As an analog modulation method, PAM is subject to serious noise limitations and thus is not normally used for long-distance communications.   PAM is sometimes utilized as a preliminary step in the generation of pulse code modulation (PCM) signals, which are used extensively in digital communications systems.

3.9 PULSE WIDTH MODULATION

Pulse width modulation (PWM) is an analog modulation method in which the width of a pulse is made proportional to the amplitude of a baseband signal $m(t)$.   A common method of generating a PWM signal is to add the modulating signal $m(t)$ to a ramp waveform and apply the sum to a comparator.   The output of the comparator will be a pulse whose duration is proportional to $m(t)$ if the reference level of the comparator is set properly.   Figure 3-25 illustrates pulse width modulation.

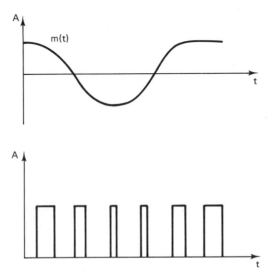

Figure 3-25 Pulse Width Modulation.

The baseband information in a PWM signal may be recovered by passing it through a low-pass baseband filter. Pulse width modulation may be time division multiplexed in a fashion identical to PAM. As with PAM, PWM is very susceptible to noise and is not normally used for long-distance communications. PWM may be utilized as a preliminary step in PCM generation.

3.10 PULSE POSITION MODULATION

Pulse position modulation (PPM) is a scheme in which the time of occurrence of a fixed-width pulse depends upon the amplitude of a sampled baseband signal $m(t)$. The baseband signal must, of course, be sampled at least at the Nyquist rate. Figure 3-26 illustrates PPM.

Pulse position modulation may be generated by first generating a pulse width modulated signal and triggering another pulse generator on the trailing edge of the PWM signal. A number of methods are available to recover the baseband information in a PPM signal, all of which require some sort of fixed phase reference. PPM signals may be time division multiplexed just as PAM signals. Pulse position modulation is also considered an analog modulation method and thus is susceptible

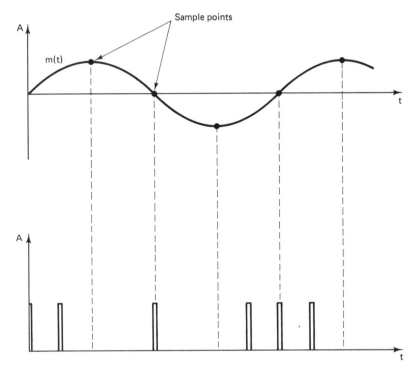

Figure 3-26   Pulse Position Modulation.

to noise, but less so than PAM because of its similarity to phase modulation. PPM, like PAM and PWM, may be utilized as a preliminary step in PCM signal generation.

EXERCISES ━━━━━━━━━━━━━━━━━━━━━━━━━━━━━━━━━━━━

1. A 5 KHz audio tone with a peak amplitude of 8.4 volts is multiplied by a 30 MHz RF signal whose peak amplitude is 25 volts. Determine the frequencies of all signals present at the multiplier output.

2. A ±0.3 volt (peak) baseband signal amplitude modulates a 1.3 GHz carrier whose amplitude is 9.3 volts (peak). Calculate the percent modulation.

3. A 2.4 GHz carrier is FM modulated by an 8 KHz sinusoidal tone with a frequency deviation of 0.8 KHz. Calculate the modulation index β of the signal.

4. Calculate the bandwidth required for a 440 MHz carrier modulated by a baseband signal band-limited to 4 KHz. The frequency deviation in this system is 1 KHz.

5. Calculate the sideband frequencies for Exercise 4.

6. Plot the relative amplitude of the carrier and all significant sidebands for a wideband FM signal with a deviation of 50 KHz and a 14.7 KHz maximum modulating frequency.

7. Generate a block diagram of a frequency division multiplexing scheme to transmit five baseband signals, each band-limited to 8 KHz over a single 14 GHz communications link.

8. Calculate the bandwidth of a 14 GHz time division multiplexed communications system if eight 4 KHz baseband signals are multiplexed onto the single carrier.

4

Analog-to-Digital and Digital-to-Analog Conversion Methods

Many baseband signals, such as voice and video, are analog in nature, while others, such as computer data streams, are digital. In telecommunications and other applications, it is often desirable to multiplex analog and digital signals together and to transmit the resulting signal through a digital communications system. Analog signals must be digitized prior to multiplexing; pulse code modulation (PCM) is the most common technique. In a PCM system the analog signal is sampled, assigned to the nearest allowed discrete voltage level (quantized), and then encoded as a series of binary pulses. A PCM system provides several such pulses (or bits) for each sample and thus requires a greater transmission bandwidth than a system such as PAM, which utilizes only a single pulse per sample. The advantage of a PCM system is that its output is digital. In a binary digital transmission scheme, the receiver needs only to know if a pulse is present or absent. In long-distance transmission, pulses may be "cleaned up" by regeneration at each repeater station. In such a system, noise does not accumulate as it does in analog links with amplifying repeaters. The quality of a pulse code modulation signal depends only upon the sampling, quantizing and encoding process, and not on the end-to-end transmission distance. Alternatives to PCM for digitizing an analog signal include delta modulation (DM) and differential pulse code modulation (DPCM).

A major advantage of any of these forms of digitization is the resulting ability to time division multiplex many signals onto a single data stream. PCM and related systems such as DM and DPCM are the very foundation of the digital transmission of analog signals.

4.1 PULSE CODE MODULATION

The first step in the generation of a PCM signal from an analog signal is sampling. The sampling theorem was discussed in Section 2.7, and sampling techniques were covered in Section 3.8 in the discussion of pulse amplitude modulation. In most PCM systems, sampling produces a PAM signal, but in some PCM systems, a pulse width modulation or pulse position modulation signal is initially generated.

Assuming a PAM sampler, a second essential building block is a holding circuit, which may be as simple as a capacitor. A holding circuit maintains the level of a PAM pulse until the circuit receives the next pulse, which is maintained until the next pulse is received, and so forth. During the holding time, the quantizer assigns a discrete level to the pulse and the encoder produces a unique multibit PCM code for that level. Figure 4-1 illustrates the minimum necessary functions in a PCM transmission system.

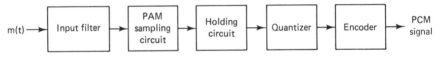

Figure 4-1 PCM transmission system.

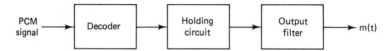

Figure 4-2 PCM receiving system.

In Figure 4-1, $m(t)$ is the baseband signal which is bandlimited to a frequency f_m by the input filter.   The PAM sampling circuit is required by the sampling theorem to sample at a rate of $2f_m$ or greater.   Figure 4-2 illustrates a PCM receiving system.   The output filter in Figure 4-2 is a low-pass filter with a bandwidth of f_m, which will recover the baseband signal from the holding circuit.

In Figure 4-3, the range of values that can be assumed by $m(t)$ is 0 volts to 1.6 volts, called the reference scale.   The discrete step size, Δv, is 0.1 volts, so the total number of steps, S, is 16 (from 0.1 volts to 1.6 volts inclusive).   Note from Figure 4-3 that the staircase approximation to $m(t)$ introduces error or quantization noise.   Quantization noise can be reduced by increasing the number of steps S, which is equivalent to decreasing the step size Δv.   A measure of quantization noise is the mean square quantizing error given by:

$$\overline{e^2(t)} = \frac{(\Delta v)^2}{12}$$

In this expression, $\overline{e^2(t)}$ is the mean square quantizing error and Δv is the step size. The rms quantizing error is the square root of the mean square quantizing error. It can be shown that all of the spectral components of the quantizing noise are concentrated within the band limits of the baseband signal.   The quantization

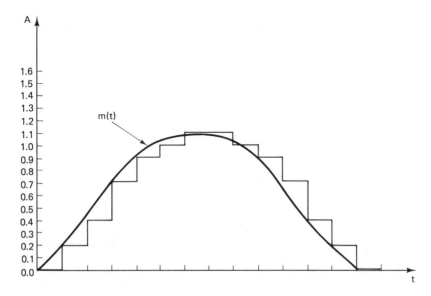

Figure 4-3 Quantization process.

illustrated in Figure 4-3 is called uniform quantization because of the fixed step size, Δv.

Recall from Section 2.5 the Hartley-Shannon law, which indicates that bandwidth may be "traded off" with signal-to-noise ratio for a given information rate. An expression illustrating this trade-off specific to a binary PCM system is as follows:

$$\frac{S_o/N_o}{S_i/N_i} = \frac{8}{K^2} 2^{\left(\frac{2B}{f_m}\right)}$$

The transmission bandwidth is dependent upon the number of bits per sample and the sampling rate. The ratio $2B/f_m$ is equal to twice the number of bits per sample. In this expression, S_o/N_o is the output signal-to-noise ratio (considering only quantization noise), S_i/N_i is the analog input signal's signal-to-noise ratio, B is the transmission bandwidth, and f_m is the maximum baseband frequency. The constant K is defined as

$$K = \frac{A}{\Delta n}$$

In this expression, Δn is the average input noise level, and A is the maximum voltage on the reference scale. The above expressions illustrate that a large improvement in the quantizing signal-to-noise ratio may be obtained by a relatively small increase in bandwidth.

Example 4-1

A 64-step binary PCM system has a reference scale from 0 to 6.4 volts. Compute the rms quantizing error for this system.

Solution

$$\Delta v = \frac{A}{S} = \frac{6.4}{64} = 0.1$$

$$\overline{e^2(t)} = \frac{(\Delta v)^2}{12}$$

The rms error is:

$$\sqrt{\overline{e^2(t)}} = \frac{\Delta v}{\sqrt{12}} = \frac{0.1}{3.46} = 0.03 \text{ volts}$$

Example 4-2

A binary PCM system has an average input noise level of 50 mv, a reference scale from 0 to 1.6 volts, and a step size Δv of 0.1 volt. The baseband signal applied to this system is band-limited to 8 KHz and the transmission bandwidth is 32 KHz.

Compute the improvement in signal-to-noise ratio from input to output for this system and compare it to the same system with a transmission bandwidth of 64 KHz.

Solution

$$K = \frac{A}{\Delta n} = \frac{1.6}{50 \times 10^{-3}} = 32$$

For $B = 32$ KHz:

$$\frac{S_o/N_o}{S_i/N_i} = \frac{8}{K^2} 2^{\left(\frac{2B}{f_m}\right)} = \frac{8}{32^2} 2^{\left(\frac{64}{8}\right)} = \frac{8}{1024} 2^8 = 2.0$$

For $B = 64$ KHz:

$$\frac{S_o/N_o}{S_i/N_i} = \frac{8}{1024} 2^{16} = 512$$

Doubling the transmission bandwidth increases the signal-to-noise ratio by a factor of 256.

The previously described quantization scheme utilized uniform quantization with a fixed value for Δv. In such a system, the quantization noise can be reduced by making Δv smaller at the expense of increased bandwidth. A noise reduction scheme that does not require increased transmission bandwidth utilizes non-uniform or non-linear quantization intervals. Non-uniform quantization takes advantage of the fact that baseband signals such as speech have a higher probability of a low amplitude than of a high amplitude. Non-linear quantization therefore provides a small Δv for low amplitudes and a larger Δv for high amplitudes. As long as the total number of steps remains the same as for the uniform quantization case, the transmission bandwidth requirements are identical. Non-uniform quantization is most often implemented by compressing the baseband signal in amplitude prior to uniform quantization and expanding it at the receiving end after decoding. This compression-expansion operation is called companding. Figure 4-4 illustrates the implementation of non-uniform quantization by companding.

Two companding laws are in common use in telecommunications. The so-called μ-law is used in North America and Japan and the A-law is used in Europe,

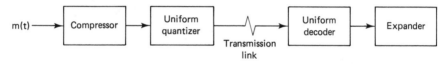

Figure 4-4 Block diagram of companding system.

Africa and South America and on most international telecommunications routes. The μ-law equation is given by:

$$V_c = \frac{a \log \dfrac{1 + \mu V_i}{a}}{\log (1 + \mu)} \quad \text{for } 0 \le V_i \le a$$

$$V_c = \frac{-a \log \dfrac{1 - \mu V_i}{a}}{\log (1 + \mu)} \quad \text{for } -a \le V_i \le 0$$

In these expressions, V_i is the input signal amplitude in volts, V_c is the compressed signal amplitude in volts, a is a parameter that defines the operating region, and μ is the compression parameter. The μ-law is normally written in normalized form ($a = 1$) and for positive values of V_i:

$$V_c = \frac{\log (1 + \mu V_i)}{\log (1 + \mu)}$$

Figure 4-5 illustrates the compression produced by the μ-law. Note that the μ-law (μ > 0) is linear for low-amplitude input signals and that it compresses high-amplitude signals. Practical values of μ are of the order of 100.

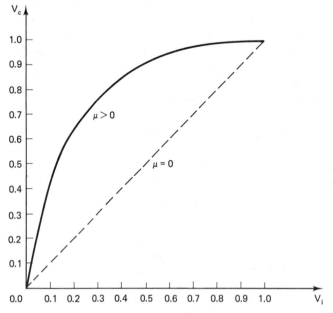

Figure 4-5 μ-law companding.

The normalized equations for the A-law are given by:

$$V_c = \frac{AV_i}{1 + \log A} \text{ for } 0 \leq V_i \leq \frac{1}{A}$$

$$V_c = \frac{1 + \log (AV_i)}{1 + \log A} \text{ for } \frac{1}{A} \leq V_i \leq 1$$

In this expression A is the compression parameter which is set at values of the order of 100 in practical systems.

Companding laws are normally approximated by line segments in practical systems, the resulting process is known as segmented companding.

Example 4-3

A normalized input signal of 0.7 is applied to a μ-law compander with $\mu = 100$ and an A-law compander with $A = 100$.   Calculate the compressed output for each system.

Solution

For the μ-law system:

$$V_c = \frac{\log (1 + \mu V_i)}{\log (1 + \mu)} = \frac{\log (1 + 70)}{\log (101)} = \frac{1.85}{2.00} = 0.93$$

For the A-law system:

$$V_c = \frac{1 + \log (AV_i)}{1 + \log A}$$

$$V_c = \frac{1 + \log (70)}{1 + \log (100)} = \frac{1 + 1.85}{1 + 2.00} = \frac{2.85}{3.00} = 0.95$$

The two systems give similar amounts of compression at this normalized input level.

The final step in the generation of a PCM signal is encoding. Encoding is the process of assigning a unique sequence of binary digits or bits to each level of quantization. The number of bits required is:

$$n = \log_2 S$$

In this expression, n is the number of bits, and S is the number of quantization steps or levels.

TABLE 4-1 BINARY CODE
FOR 8-LEVEL PCM

Quantization Level	Binary Code
1	000
2	001
3	010
4	011
5	100
6	101
7	110
8	111

Example 4-4

Compute the number of bits necessary to encode a 16-level PCM system.

Solution

$$n = \log_2 S = \log_2 16 = 4$$

Each quantization level would be given a unique 4-bit code.

Although any number of different coding schemes is possible in a PCM system, two of the most used include the straight binary code and the Gray code. The binary code is given in Table 4-1 for an 8-level PCM system, and the Gray code is illustrated in Table 4-2 for the same 8-level system. The advantage of the Gray code is that the code changes by exactly one bit for each change in level. This can minimize potential errors in the decoding of the PCM signal.

PCM encoders are sometimes designed with more bits than necessary for the number of levels of quantization. Excess bits may be used for error detection and

TABLE 4-2 GRAY CODE
FOR 8-LEVEL PCM

Quantization Level	Gray Code
1	000
2	001
3	011
4	010
5	110
6	111
7	101
8	100

correction.   Transmission of a parity bit is an example of an error-detection scheme. Error-control techniques are the subject of Chapter 5.

4.2 DELTA MODULATION

Delta modulation is a form of analog-to-digital conversion in which the change of the current quantization level from the previous quantization level is encoded rather than the absolute value of the current level.   In delta modulation, the maximum change that may occur for each sampling instant is one quantum level.   Figure 4-6 illustrates the concept of quantization using delta modulation.   Figure 4-6 illustrates that the behavior of $m(t)$ is communicated by transmitting a positive pulse when $m(t)$ has increased since the last sampling instant and a negative pulse when $m(t)$ has decreased.   When $m(t)$ remains at a fixed level, the system transmits alternate negative and positive pulses.

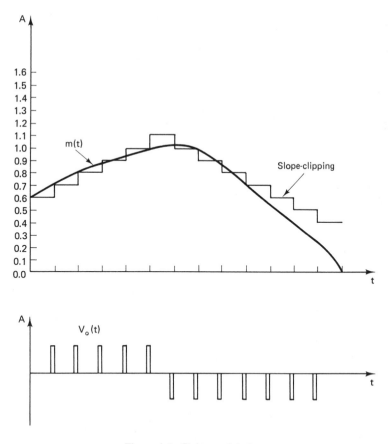

Figure 4-6   Delta modulation.

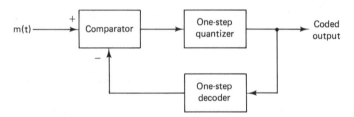

Figure 4-7   Delta modulation generation.

An advantage of a delta modulation system is that it can follow all amplitude variations, so that peak-clipping is not a problem.   One of its disadvantages is that it may be unable to follow rapidly changing signals and thus is subject to a problem known as slope-clipping.   Slope-clipping may be combatted by increasing the sampling rate and/or decreasing the step size.   Practical DM systems sample at approximately ten times the minimum Nyquist rate.   Figure 4-7 illustrates the system required for the generation of delta modulation.   A delta modulation coded signal may be recovered with a one-step decoder followed by an integrator and low-pass filter.   Figure 4-7 indicates that a delta modulation system is much simpler to implement than a PCM system.   The disadvantage of DM as compared to PCM is that one quantizer is required for each channel as opposed to PCM where a single quantizer may be utilized to encode a number of multiplexed channels.   Delta modulation is often utilized when a single channel is to be digitized.   Integrated circuit codec (coder-decoder) chips often use DM.

4.3 DIFFERENTIAL PULSE CODE MODULATION

In differential pulse code modulation (DPCM), the difference between successive samples is encoded as several binary digits rather than as a single digit as in delta modulation.   This method of digital-to-analog conversion is well suited to signals that do not change rapidly (i.e., where the correlation between successive samples is high).   For such signals, DPCM provides an increased signal-to-noise ratio over PCM for the same number of bits per sample.   The price paid for this improvement is the increased circuit complexity required to encode the difference signal into several digits.

A variation of DPCM is based upon the fact that the message signal carries redundant information, since successive samples are highly correlated.   Predictive quantization is a technique where redundancy is effectively removed by predicting the value of the next sample based upon the values of previous samples.   The system then encodes the difference between the predicted and actual values.   Predictive quantization is more complex to implement than DPCM or PCM but allows significantly reduced data rates for the same information throughput.

4.4 PULSE CODE MODULATION TIME DIVISION MULTIPLEXING ─────

Time division multiplexing (TDM) was introduced in Section 3.7 as a system able to combine pulse amplitude modulation, pulse width modulation, or pulse position modulation signals into a single communications channel. Although this application of TDM is important, its greatest use is with PCM systems. Digital time division multiplex systems may be classified as synchronous or asynchronous. In a synchronous TDM system, the data streams are all derived from the same clock. In an asynchronous TDM system, the data streams are derived from different clock sources. An asynchronous TDM system synchronizes the data streams to a common clock by a process known as justification or pulse stuffing; then, the signals are multiplexed by a synchronous TDM system. This section treats only synchronous systems.

The serializer is the fundamental subsystem in a TDM system. It takes the PCM words from the channels to be multiplexed and merges them into a single serial data stream. A serializer may produce word-interleaved or bit-interleaved data streams. In word-interleaved systems, the technique most commonly used in telecommunications systems, the time slot allocated to each channel sampled is long enough to accommodate a complete PCM word. Since all signals at input are digital, a serializer is easily implemented with a highly stable master clock signal and logic gates. Serializers are available in integrated circuit form or may be implemented by software in microprocessor-based systems.

A multiplexer requires some method to indicate to the demultiplexer at the receiving end which time slot is associated with which channel. This is normally accomplished by interleaving a predetermined binary synchronization word with the data words. This word is known as a frame alignment word (FAW); the frame alignment word plus all of the data words from each channel form a frame. Sometimes two or more frame alignment words are used. Frame alignment words may be "bunched" at the start of a frame or "distributed" within the frame. Frame alignment words may be detected with a shift register and logic gates or with software in a microprocessor-based system. TDM systems utilized in telecommunications applications also require "housekeeping bits" for network supervision; these bits are normally spare or otherwise unused bits but sometimes they are "stolen" from a data word. The multiplexed output data rate is given by:

$$f_o = wf_s(c + n)$$

In this expression, f_o is the output data rate, w is the number of digits used to represent each coded sample, f_s is the sampling frequency, c is the number of channels, and n is the number of frame alignment words.

Example 4-5

An 8-bit PCM system samples 24 voice channels, compresses them using the μ-law, samples them at a rate of 8 KHz, and utilizes one frame alignment word in a time division multiplexed system.   Compute the output date rate of the TDM system.

Solution

$$f_o = wf_s(c + n) = (8)(8 \times 10^3)\ (24 + 1)$$

$$f_o = 1.6 \times 10^6 = 1.6\ \text{Mbit/sec}$$

PCM-TDM systems are widely used in telecommunications and will be discussed further in Chapter 19.

4.5 BASEBAND PULSE CODE MODULATION TRANSMISSION

A PCM signal is rarely transmitted directly over long distances.   It is usually applied to a digital carrier modulator in the case of transmission via a radio or satellite link or must be coded into a different form for transmission over electrical or optical cables such as those found in telecommunications systems.   Chapter 6 is devoted to digital modulation methods; the remainder of this section will briefly discuss the methods used for baseband cable transmission.

The code to be transmitted through the cable should have a constant DC level along the transmission line, minimal energy at lower frequencies, and the ability to allow the extraction of timing data.   The simplest format that satisfies these criteria is the alternate mark inversion (AMI) format, which provides alternate positive and negative pulses for a logical 1 and zero volts for a logical 0. Figure 4-8 illustrates a nonreturn-to-zero (NRZ) AMI formatted signal.   Note that the transmitted "marks" (positive or negative pulses) occupy a full clock period in a nonreturn-to-zero coding scheme. Figure 4-9 illustrates a return-to-zero (RZ) format scheme. In the return-to-zero formatting scheme illustrated in Figure 4-9 the mark pulses occupy one-half of the clock period, so that the mark/space ratio is 50%.   Although the mark-space ratio is normally 50% in metallic cable systems, it is often reduced to 30% or less in optical systems.

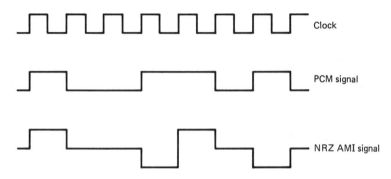

Figure 4-8   NRZ AMI formatting of a PCM signal.

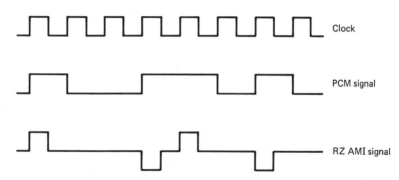

Figure 4-9 RZ AMI formatting of a PCM signal.

A constant DC level along the transmission line is important to simplify the circuitry that detects the presence of a mark as well as the automatic gain control circuitry within a repeater or regenerator. The most important reason for a constant DC level, however, is that the transmission line is normally used to provide DC power to the repeaters in addition to carrying the PCM signal; thus variations in the DC level would create serious difficulties.

A format that produces a minimal number of frequency components at the lower end of the spectrum is important, because cables often include equalization circuits to compensate for frequency-dependent cable parameters and pulse distortion. A significant low-frequency component would require physically large equalization components. An additional reason for minimizing low-frequency components is the fact that cable attenuation decreases as the frequency is decreased, and it is possible that the low-frequency components could overload the repeaters if they were not reduced.

The extraction of timing information is important for synchronization at the receiving end. Timing extraction is facilitated by a format with numerous zero crossings. A disadvantage of AMI formatting is that timing information (zero crossings) is not available if the PCM signal consists of a long sequence of logical zeroes. Various techniques have been developed to overcome this problem, including inversion of even-numbered bits within each sample prior to multiplexing. More complex formats are also in use that limit the number of consecutive zeroes (or ones) that may be transmitted.

4.6 ERROR PERFORMANCE OF BASEBAND PULSE CODE MODULATION SYSTEMS

The receiver in a pulse code modulation system is required to make a decision on each bit of data to determine if a logical 1 or 0 was sent. By the time the transmitted PCM signal is received, it has been corrupted by noise. Noise may be randomly distributed in time or may occur in bursts. The effects of random noise on PCM system performance will be discussed here. The effects of burst noise are more difficult to quantify; because of its unpredictable nature, it will not be discussed further.

The type of noise to be considered here is called Additive White Gaussian Noise (AWGN). Additive simply indicates that the noise encountered along the transmission path is added to the signal. White indicates that the noise is frequency independent at least over the spectral range of interest, and Gaussian indicates that the noise voltage may be described by a bell-shaped curve known as a normal or Gaussian distribution. AWGN has a two-sided (both + and − frequency) spectral density that is a constant value of $N_o/2$. N_o has units of volts^2/Hz.

The error performance of a digital communications system is described by a quantity called the probability of error, P_e, also called the bit error rate (BER). P_e depends upon the ratio of signal energy in a bit to the quantity N_0. For a bipolar pulse, the average bit signal energy is given by

$$E_b = V^2 T$$

In this expression, E_b is the average bit signal energy, V is the pulse amplitude, and T is the pulse duration. Bipolar pulses represent a logical 1 by $+V$ and logical 0 by $-V$ or vice versa. The probability of error for a baseband PCM system with a so-called "matched filter" at the receiver is

$$P_e = \frac{1}{2} \, erfc \left(\sqrt{\frac{E_b}{N_0}} \right)$$

In this expression, *erfc* represents a rather complex mathematical function known as the complementary error function. This function is tabulated in various handbooks, and selected values of *erfc* are given in Table 4-3.

A matched filter is the one filter that maximizes the signal-to-noise ratio for a given waveform. Discussion of matched filter design is beyond the scope of this

TABLE 4-3 SELECTED VALUES OF THE COMPLEMENTARY ERROR FUNCTION

x	erfc (x)
0.0	1.0
0.4	5.72×10^{-1}
1.0	1.57×10^{-1}
1.4	4.87×10^{-2}
2.0	7.21×10^{-3}
2.4	6.9×10^{-4}
3.0	2.3×10^{-5}
4.0	1.5×10^{-8}
5.0	1.5×10^{-12}

For $a > 2$, $erfc(a) \approx \dfrac{e^{-a^2}}{a\sqrt{\pi}}$

text, but it may be noted that the matched filter for the baseband PCM system described here is an integrator circuit.

Reasonable values of P_e in a digital communications system range from 10^{-4} which is rather poor, but acceptable for speech signals, to 10^{-9} or better.

Example 4-6

A baseband PCM system with a matched filter at the receiver is implemented with ± 5 volt bipolar pulses with a pulse duration of 72 μsec. If the noise spectral density is 1.0×10^{-4} volts²/Hz, calculate the probability of error for this system.

Solution

$$E_b = V^2 T_b = (5)^2 (72 \times 10^{-6}) = 1.8 \times 10^{-3}$$

$$\frac{N_o}{2} = 1.0 \times 10^{-4} \qquad N_o = 2 \times 10^{-4}$$

$$P_e = \frac{1}{2} erfc \left(\sqrt{\frac{E_b}{N_o}} \right) = \frac{1}{2} erfc \left(\sqrt{\frac{1.8 \times 10^{-3}}{2 \times 10^{-4}}} \right)$$

$$P_e = \frac{1}{2} erfc (3) = \frac{2.3 \times 10^{-5}}{2} = 1.15 \times 10^{-5}$$

It should be noted that average bit energy would be computed differently for an NRZ or RZ AMI coded PCM signal. The given expression for T_b assumes a bipolar pulse and that ones and zeroes are equally likely.

Probability of error (bit error rate) will also be used to compare the various methods of digital modulation discussed in Chapter 6.

EXERCISES

1. A 32-step binary PCM system has a reference scale from 0 to 9.6 volts. Calculate the rms quantizing error for this system.
2. A 64-step binary PCM system has a reference scale from 0 to 6.4 volts and an average input noise level of 100 mv. A band-limited (5 KHz) baseband signal is transmitted with a transmission bandwidth of 30 KHz. Calculate the improvement in signal-to-noise ratio from input to output for this system.
3. Calculate the improvement in signal-to-noise ratio in Exercise 2 if the transmission bandwidth is increased to 40 KHz.
4. A normalized input signal of amplitude 0.1 is applied to a μ-law compander with μ = 75. Calculate the normalized output of the compander.

5. Calculate the number of levels in a PCM system utilizing a 5-bit code.

6. An A-law, 16-channel, 12-bit PCM system samples at 8 KHz and utilizes two frame alignment words in a TDM system. Compute the output data rate of this system.

7. Estimate *erfc* (1.2) and *erfc* (4.5).

8. A baseband PCM system with a matched filter at the receiver is implemented with ± 12 volt pulses of duration 50 μsec. If the noise spectral density is 1.5 $\times$ 10^{-4} volts^2/Hz, estimate the error probability for this system.

5

Error Control Codes

Digital communications systems transfer information between points by means of one or more streams of binary digits. The bit streams may have originated in a computer or other digital device, or they may have been generated from an analog signal by an analog-to-digital conversion technique, such as PCM. The bit streams may be transmitted at baseband or may utilize one of the modulation techniques to be discussed in Chapter 6 to modulate a carrier. In either case, the data streams will be subject to noise. The following discussion will consider a single bit stream of serial data where a sequence of bits represents one data character or word. The noise encountered on a bit stream may be random, such as thermal noise, or it may occur in bursts, perhaps caused by lightning or other electrical interference. A portion of the data stream occasionally may be lost due to signal fading. These realistic conditions on communications channels can cause logical ones to be interpreted as logical zeros and logical zeros to be interpreted as logical ones. The result is a transmission that contains bit errors. The effect of these errors on a digital communications system ranges from distortion on a digital voice system to erroneous data transmission in a computer communications system.

By a suitable implementation of redundant coding and/or retransmission techniques, it is possible to reduce the effective errors in digital communications systems to negligible proportions. Error control in a digital communications system consists of error detection and some method for the correction of the detected errors.

5.1 ERROR CONTROL STRATEGIES

In general, digital communications systems may either be one-way, with a single transmitter and one or more receivers, or two-way, with receivers and transmitters at each location. In one-way systems, error control must be accomplished by using *forward error correction* (FEC) techniques. In a communications system with FEC, errors introduced by the communications channel are detected and corrected at the receiver. A deep-space probe is a one-way communications system that uses FEC. It sends data to the earth over a noisy communications channel, which introduces random errors. These random errors are detected and corrected at the receiver on the earth. Forward error correction is also used when computers write data to magnetic media such as tape or disk, as well as in digital magnetic and optical audio and video recording systems.

Most digital communications systems are two-way. Examples of two-way digital communications links are digital transmission over the public telecommunications networks and digital satellite two-way links. Error control in two-way systems is often accomplished using *automatic repeat request* (ARQ) systems. ARQ systems are especially effective in combatting the effects of burst errors. ARQ is used in packet-switched telecommunications networks, packet radio and other data communications systems.   In a two-way communications system utilizing ARQ, errors introduced into a data block by the communications channel are detected at the receiver, which requests the sender to retransmit any erroneous blocks. Retransmission requests will continue until the block is received correctly or the system terminates requests after a finite number of tries.   The two types of ARQ are *stop-and-wait* ARQ and *continuous* ARQ.   In stop-and-wait ARQ, the transmitter sends out a block of data and waits for information from the receiver.   If errors are detected by the receiver, it requests a repeat of the erroneous block of data.   If no errors are detected, the receiver requests the next data block from the transmitter.   In continuous ARQ, the transmitter sends data blocks to the receiver continuously.   If the receiver detects an error in a data block, it requests a repeat.   If the erroneous data block and all succeeding data blocks are transmitted again, even if they contain no errors, then the system is known as a *go-back*-N or *pipelined* continuous ARQ system.   An alternate approach is for the transmitter to resend only erroneous data blocks; this is known as *selective repeat* continuous ARQ.   Continuous ARQ is more efficient than stop-and-wait ARQ, and selective repeat continuous ARQ is more efficient than go-back-*N* continuous ARQ.   The price to be paid for this increased efficiency is additional complexity and increased memory requirements in the digital circuitry required to implement the ARQ system.

In general, ARQ systems require less complex digital circuitry than FEC systems and are generally preferred.   If the error rate of the communications channel is high, however, retransmissions are frequent and the system throughput suffers.   In this situation, a hybrid FEC/ARQ system may be the most efficient approach.

5.2 BLOCK CODES

In digital communications systems, an information data stream is normally partitioned into sequences of binary digits called *words*.   Groups of words are called *blocks*.   If additional redundant bits are included with the information words and/ or blocks, the resulting string of additional binary digits is said to be a *block code*. By increasing the number of redundant bits, the probability of undetected error in a digital communications system may be made arbitrarily small. The trade-off for this reduction in error probability is increased overhead (additional non-information-carrying bits), which reduces system throughput.   Block codes may be used to detect errors for ARQ applications or to detect and correct errors in FEC applications.

The most common block code is generated by adding a single redundant bit to each information word. This redundant bit, called a parity bit, is appended to each information word, so that the total number of ones (including the parity bit) is odd for odd parity or even for even parity. The parity bit is inserted at the transmitter, and even or odd parity is checked at the receiver. The receiver must know in advance whether even or odd parity is in use. Systems using a single parity bit assume that the most likely error is a single bit error, since the technique will not detect a two-bit error or an even number of bit errors per word.

Example 5-1

Compute the value of the even parity bit P for the following words: $P11001110$, $P00010010$, $P11100001$. Assume that the words are transmitted LSB first (i.e., read them from right to left).

Solution

In even parity, the total number of ones in the code must be even, so the words with the parity bit appended at position P are:

$$111001110, 000010010, 011100001$$

Example 5-2

The following words are received in an odd parity system: 100011010, 010001001, 111000110. Which of these words, if any, were received erroneously?

Solution

100011010 has an even number of ones, so it is in error.

01001001 has an odd number of ones, so at least it does not have an odd number of bits in error.

111000110 has an odd number of ones, so it also does not have an odd number of bits in error.

If two parity bits are used, it is possible to detect and correct single bit errors. The most common technique for single bit error detection and correction is to use a parity bit for each word and a block check sequence for each block of words. The block check sequence is formed by having as the first bit in the sequence a parity bit based upon the first bit of each word in the block, the second bit in the sequence based upon the second bit in each word and so on. This technique is sometimes called a *vertical redundancy check* or VRC (word parity bits) and a *longitudinal redundancy check* or LRC (block check sequence).

Example 5-3

The following digital words form a block: 00110011, 11001100, 10101010, 01010101, 11100011, 00010100, 10011011, 11111110. Compute the odd parity sequence and even parity block check sequence for this data stream. Each word is transmitted LSB first.

Solution

Arrange the data words in block format from LSB to MSB vertically, and calculate the parity sequence and block check sequence:

	Word 1 2 3 4 5 6 7 8	Block check sequence word
LSB	1 0 0 1 1 0 1 0	0
	1 0 1 0 1 0 1 1	1
	0 1 0 1 0 1 0 1	0
	0 1 1 0 0 0 1 1	0
	1 0 0 1 0 1 1 1	1
	1 0 1 0 1 0 0 1	0
	0 1 0 1 1 0 0 1	0
MSB	0 1 1 0 1 0 1 1	1
Word parity bits	1 1 1 1 0 1 0 0	0

The coded data stream (with the parity bit sent following the MSB of each word) would be 100110011, 111001100, 110101010, 101010101, 011100011, 100010100, 010011011, 011111110, 010010010. Note that the last coded word is the block check word. Note also that the first bit (last transmitted bit) of the block check word is a parity bit computed as if the block check word was a data word (odd parity in this example).

Example 5-4

An error occurs in the LSB of the third word in the data block of Example 5-3. Describe how the error is detected and corrected by the code.

Solution

The received block (with word 9 being the BCS word) is

	Word
	1 2 3 4 5 6 7 8 9

		1 2 3 4 5 6 7 8 9	
	1–	1 0 1 1 1 0 1 0 0	LSB
	2–	1 0 1 0 1 0 1 1 1	
	3–	0 1 0 1 0 1 0 1 0	
	4–	0 1 1 0 0 0 1 1 0	
Bit	5–	1 0 0 1 0 1 1 1 1	
positions	6–	1 0 1 0 1 0 0 1 0	
	7–	0 1 0 1 1 0 0 1 0	
	8–	0 1 1 0 1 0 1 1 1	MSB
	9–	1 1 1 1 0 1 0 0 0	parity

Note that the data words are arranged vertically from LSB to MSB.

The receiver would know an error has occurred by noting that each word should be odd parity, and yet the third word has an even number of ones, so at least one bit may be in error. The block check sequence is computed by using even parity, so the receiver knows that an error has occurred in the LSB of a word (odd number of ones). The intersection of the third word and the LSB determines the bit in error. Once the receiver detects the location of the error, it can correct it by simply inverting the bit.

A widely used technique for FEC systems is error detection and correction with *Hamming codes*, first described by Richard Hamming in 1950. Hamming codes perform multiple parity checks on each data word but do not utilize block information. The number of redundant bits required in a Hamming code is given by:

$$2^n \geq m + n + 1$$

In this expression, n is the number of redundant bits and m is the number of bits in the input data word. The expression is solved by trial and error.

Example 5-5

Calculate the number of redundant bits required if a Hamming code is to be used with an 8-bit data word.

Solution

$$m = 8$$

$$2^n \geq m + n + 1$$

Try $n = 3$:

$$2^3 \overset{?}{\geq} 8 + 3 + 1 \qquad 8 \overset{?}{\geq} 12$$

Three bits are insufficient.

Try $n = 4$:

$$2^4 \overset{?}{\geq} 8 + 4 + 1 \qquad 16 \overset{?}{\geq} 13$$

Four bits are sufficient.
The Hamming code would be a 12-bit transmitted data word (or code word) for each 8-bit data word.

The redundant Hamming bits may be placed anywhere in the data string, but, of course, the receiver must know where to expect them. Locating them properly will optimize the detection and/or correction of certain types of burst errors. The values of the Hamming bits (1 or 0) are determined by expressing the position of each bit in the code word (Hamming bits plus data bits) as an n-bit binary number and performing the exclusive-or (XOR) logical operation on all positions containing a logical one. At the receiver, the Hamming bits are extracted and XORed with all bit positions containing a one. The result gives the position of any single-bit error. Additional Hamming bits allow detection and correction of multiple-bit errors.

Example 5-6

Encode the following 8-bit data word with appropriate Hamming bits for error detection and correction: 11001100.

Solution

Four Hamming bits are required, which may be placed anywhere in the data stream. The positions with the letter H in the data stream below indicate one possible arrangement:

12	11	10	9	8	7	6	5	4	3	2	1
H	1	1	H	0	0	H	1	1	H	0	0

Note that the bit positions are indicated by numbers and that logical ones are present in positions 4, 5, 10 and 11. The 4-bit binary numbers representing these bit positions are 0100, 0101, 1010, and 1011. Performing the exclusive-or on the first set of numbers:

$$
\begin{array}{r}
0\ 1\ 0\ 0 \\
\oplus 0\ 1\ 0\ 1 \\
\hline
0\ 0\ 0\ 1
\end{array}
$$

The above result is XORed with the next number:

$$
\begin{array}{r}
0\ 0\ 0\ 1\\
\oplus 1\ 0\ 1\ 0\\
\hline
1\ 0\ 1\ 1
\end{array}
$$

The above result is XORed with the last number:

$$
\begin{array}{r}
1\ 0\ 1\ 1\\
\oplus 1\ 0\ 1\ 1\\
\hline
0\ 0\ 0\ 0
\end{array}
$$

The above four bits are the Hamming bits that replace the letter H starting with the MSB. The final Hamming code for the given data word is:

```
H     H     H     H
0  1  1  0  0  0  0  1  1  0  0  0
```

Example 5-7

The Hamming code of Example 5-6 is transmitted over a data link and the following code is received: 011000011010. Illustrate how the receiver would detect and correct the error.

Solution

The receiver extracts the Hamming bits and XORs them with the binary number corresponding to all bit positions containing a logical one:

$$
\begin{array}{ll}
\text{Hamming bits:} & 0\ 0\ 0\ 0\\
\text{Position 11:} & \underline{\oplus 1\ 0\ 1\ 1}\\
 & 1\ 0\ 1\ 1
\end{array}
$$

$$
\begin{array}{ll}
\text{Position 10:} & \underline{\oplus 1\ 0\ 1\ 0}\\
 & 0\ 0\ 0\ 1
\end{array}
$$

$$
\begin{array}{ll}
\text{Position 5:} & \underline{\oplus 0\ 1\ 0\ 1}\\
 & 0\ 1\ 0\ 0
\end{array}
$$

$$
\begin{array}{ll}
\text{Position 4:} & \underline{\oplus 0\ 1\ 0\ 0}\\
 & 0\ 0\ 0\ 0
\end{array}
$$

$$
\begin{array}{ll}
\text{Position 2:} & \underline{\oplus 0\ 0\ 1\ 0}\\
 & 0\ 0\ 1\ 0
\end{array}
$$

The result 0010 = 2 is the bit position of the error, which would be corrected by inversion.

It should be noted that the scheme illustrated in Examples 5-6 and 5-7 will not identify the bit that is in error correctly if it is one of the Hamming bits.

Integrated circuits are available that utilize Hamming codes for error detection and correction. An example is the Intel 8206, which detects and corrects all single-bit errors and can detect all double-bit errors and some multiple-bit errors.

A major disadvantage of parity bits is the overhead associated with the additional bits. An alternate approach is the checksum, which detects erroneous data blocks and thus is used only in ARQ systems. The checksum is a word sent at the end of a block of data, computed from the other words in the block. One method for generating a checksum word is by summing all of the words in the block, ignoring any additional left-end carry bits. The twos complement of the sum (complement all bits and add 1) becomes the checksum word that is sent by the transmitter. At the receiver, all the received words are summed (including the checksum word) with the additional left-end bits generated by carries ignored. If the sum at the receiver is zero, the probability that the block was received correctly is very high.

Example 5-8

Calculate the checksum word for the following block of data: 10101100, 11100111, 00110000, 10101010.

Solution

Add the data words and ignore carries:

```
                    0 0   1 0 1 0 1 1 0 0
                    0 0   1 1 1 0 0 1 1 1
                    0 0   0 0 1 1 0 0 0 0
                    0 0   1 0 1 0 1 0 1 0
                    ─────────────────────
   (ignore carries) 1 0   0 1 1 0 1 1 0 1
```

Complement of sum: 10010010
Twos complement of sum: 10010011
The checksum is 10010011.

Example 5-9

A block consisting of the four data words of Example 5-8 and the checksum is received. Illustrate how the receiver would check the block for errors.

Solution

Sum the received words in the block (ignore carries):

```
0 0   1 0 1 0 1 1 0 0
0 0   1 1 1 0 0 1 1 1
0 0   0 0 1 1 0 0 0 0
0 0   1 0 1 0 1 0 1 0
0 0   1 0 0 1 0 0 1 1   (checksum)
```
(ignore carries)   1 1   0 0 0 0 0 0 0 0

Since the sum (with carries ignored) equals zero, the block was received correctly.

The checksum technique is especially effective in combatting burst errors, since it depends upon the entire data block.

5.3 CYCLIC CODES

Cyclic codes have the property that a cyclic shift of the code word results in another code word. The most often used cyclic code is the *cyclic redundancy check* (CRC). The CRC is a set of binary digits appended to a data stream for error-checking purposes. The CRC digits are sometimes called the block check sequence (BCS) or frame check sequence (FCS).

The CRC is generated by considering the data stream bits as coefficients of a *message polynomial $M(x)$* of order $n - 1$, where n is the number of binary digits in the data stream. The least significant binary digit becomes the coefficient of the highest order polynomial term if the LSB is sent first.

Example 5-10

Form the message polynomial for the 16-bit data stream 0110011010110011.

Solution

$$M(x) = 1 \cdot x^{15} + 1 \cdot x^{14} + 0 \cdot x^{13} + 0 \cdot x^{12} + 1 \cdot x^{11} + 1 \cdot x^{10}$$
$$+ 0 \cdot x^9 + 1 \cdot x^8 + 0 \cdot x^7 + 1 \cdot x^6 + 1 \cdot x^5 + 0 \cdot x^4$$
$$+ 0 \cdot x^3 + 1 \cdot x^2 + 1 \cdot x^1 + 0 \cdot x^0$$
$$M(x) = x^{15} + x^{14} + x^{11} + x^{10} + x^8 + x^6 + x^5 + x^2 + x$$

The CRC bits are formed by dividing the message polynomial by a *generator polynomial*, $G(x)$.   When this division is performed, the result will be a quotient and a remainder.   The quotient is discarded and the remainder (which is one order shorter than $G(x)$) is used to form the CRC bits that are appended to the data stream.   At the receiver, the received data stream (with the CRC bits appended) is again divided by the generator polynomial; if this division results in a remainder of zero, the block was received correctly.   Two widely used generator polynomials are

$$G(x) = x^{16} + x^{15} + x^2 + 1$$

and

$$G(x) = x^{16} + x^{12} + x^5 + 1$$

Example 5-11

Form the CRC bits for the 8-bit data block 10111011 using $x^5 + x^4 + x + 1$ as a generator polynomial.

Solution

Form the message polynomial:

$$10111011 \rightarrow x^7 + x^6 + x^4 + x^3 + x^2 + 1$$

Divide by $G(x)$:

$$
\require{enclose}
\begin{array}{r}
x^2 \\
x^5 + x^4 + x + 1 \enclose{longdiv}{x^7 + x^6 + x^4 + x^3 + x^2 + 1} \\
\underline{x^7 + x^6 \quad x^3 + x^2 } \\
x^4 + 1
\end{array}
$$

The remainder is $x^4 + 1$, which corresponds to 10001.
The data block with CRC bits is 101110110001.

Example 5-12

The data block and CRC bits from Example 5-11 are received as sent.   Show the calculation performed at the receiver.

Solution

$$1011101110001 \longrightarrow x^{12} + x^8 + x^7 + x^6 + x^4 + x^3 + x^2 + 1$$

$$
\begin{array}{r}
x^7 + x^6 + x^5 + x^4 + x^3 \qquad\quad + x + 1 \\
x^5 + x^4 + x + 1\,\overline{)\,x^{12} \qquad\qquad\qquad + x^8 + x^7 + x^6 \qquad + x^4 + x^3 + x^2 \qquad + 1} \\
\underline{x^{12} + x^{11} \qquad\qquad\qquad + x^8 + x^7} \\
x^{11} \qquad\qquad\qquad\qquad + x^6 \qquad + x^4 + x^3 + x^2 \qquad + 1 \\
\underline{x^{11} + x^{10} \qquad\qquad\qquad + x^7 + x^6} \\
x^{10} \qquad\qquad\qquad + x^7 \qquad\qquad + x^4 + x^3 + x^2 \qquad + 1 \\
\underline{x^{10} + x^9 \qquad\qquad\qquad + x^6 + x^5} \\
x^9 \qquad\quad + x^7 + x^6 + x^5 + x^4 + x^3 + x^2 \qquad + 1 \\
\underline{x^9 + x^8 \qquad\qquad\qquad + x^5 + x^4} \\
x^8 + x^7 + x^6 \qquad\qquad + x^3 + x^2 \qquad + 1 \\
\underline{x^8 + x^7 \qquad\qquad\qquad + x^4 + x^3} \\
x^6 \qquad + x^4 \qquad + x^2 \qquad + 1 \\
\underline{x^6 + x^5 \qquad\qquad\qquad + x^2 + x} \\
x^5 + x^4 \qquad\qquad + x + 1 \\
\underline{x^5 + x^4 \qquad\qquad + x + 1} \\
0
\end{array}
$$

Note that whenever the result of a "subtraction" was negative, the absolute value of that term was used. This is equivalent to forming the exclusive-or of the coefficients of each term involved.

The remainder of zero indicates that the data block and CRC bits were received correctly, as the example stated.

CRC calculations may be done in software or they may be implemented in hardware with shift registers and logic gates. Since the CRC check only detects errors, it is appropriate only for ARQ error control systems, not for forward error correction systems.

5.4 CONVOLUTIONAL CODES

Convolutional codes differ from block codes in that the code generated depends not only upon the information data stream at a given point in time, but also upon the past history of that data stream. Convolutional codes have been developed to detect and correct random errors, burst errors or both types. Convolutional codes have been shown to be equal or superior to block codes in practical error control applications. A convolutional code is normally generated by passing the

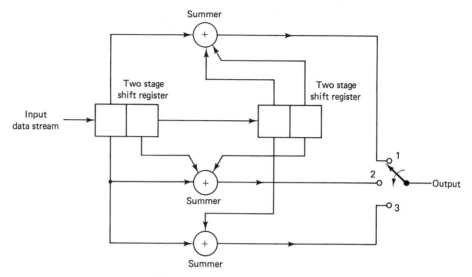

Figure 5-1 Convolutional encoder.

information data stream through a multi-stage shift register and algebraically combining the outputs to form the encoded sequence. Figure 5-1 illustrates a simple convolutional encoder where two data stream bits at a time are moved into a two-stage shift register, and three output bits are generated by algebraically combining the shift register outputs.

Further discussion of convolutional codes and decoding techniques is beyond the scope of this text. The interested reader is referred to more advanced texts on error-control codes.

EXERCISES

1. Compute the value of the odd parity bit P for the following words: $P0110011$, $P1111010$, $P0000011$, $P1111111$.

2. The following digital words form a block: 10101100, 11001100, 00110000, 10101010, 00001111, 11110000, 11111111, 10010010. Compute the even parity sequence and block check sequence for this data stream. Read each word from right to left.

3. An error occurs in the 4th bit (from the right) of the 6th word of the data block of Exercise 2. Describe how the error is detected and corrected by the code.

4. Encode the following 16-bit data word with appropriate Hamming bits for error detection and correction: 1011001110111001.

5. An error occurs in the MSB of the Hamming code of Exercise 4. Illustrate how the receiver would detect and correct the error.

6. Calculate the checksum word for the data block in Exercise 2 using the algorithm described in this chapter.

7. A block consisting of the data words of Exercise 2 and the checksum of Exercise 6 is received. Illustrate how the receiver would check the block for errors.

8. Form the CRC bits for the 8-bit data block 01110011 using $x^5 + x^4 + x + 1$ as the generator polynomial. Read the data string from right to left.

9. The data block and the CRC bits from Exercise 8 are received as sent. Illustrate how the receiver would check the block for errors.

6

Digital Modulation Methods

Although transmission of PCM signals at baseband over short distances has numerous applications, most digital communications systems use the baseband PCM signal to modulate a carrier. A number of digital modulation methods are available to the communications system designer. The choice between methods is made on the basis of the relative power required or the bandwidth necessary for a given probability of error or bit error rate performance. Relative circuit complexity may also be a consideration. The modulation methods to be discussed in this chapter include the most popular techniques, but the reader should be aware of the fact that additional, less-known methods do exist and may become important in the future.

6.1 FREQUENCY SHIFT KEYING

One of the most straightforward and widely used forms of digital modulation is *frequency shift keying* (FSK). FSK has been used for many years in radioteletype applications and is the technique used in low-speed modems such as the 300 baud units utilized in personal computer communications over dial-up telecommunications lines. In a frequency shift keying system, a binary bit stream such as a PCM signal is used to generate the following waveform:

$$v(t) = A \cos (\omega_c \pm \Delta\omega)t$$

In this expression, A is the amplitude of the FSK waveform, ω_c is the carrier frequency, $\Delta\omega$ is the frequency deviation, and t is time. The $+$ or $-$ is a function of the sense of the bit (logical 1 or logical 0). An FSK signal has a constant amplitude and one of two carrier frequencies, $\omega_c + \Delta\omega$ or $\omega_c - \Delta\omega$.

An FSK signal may be demodulated either *synchronously* (coherently) or *nonsynchronously* (noncoherently). Coherent demodulation is possible only with coherent FSK signals (i.e., those generated by frequency modulating a single oscillator). Non-coherent FSK is generated by switching between two free-running oscillators at the two frequencies $\omega_c \pm \Delta\omega$. Coherent demodulation makes use of the phase information in the FSK signal and performs better in the presence of noise than does noncoherent demodulation, which makes use of only amplitude information. Noncoherent demodulation is extremely simple in that it requires only that the received signal be applied to two narrowband bandpass filters, one tuned to $\omega_c + \Delta\omega$ and the other to $\omega_c - \Delta\omega$. The filter with the larger output indicates whether a logical one or logical zero has been sent. Figure 6-1 illustrates noncoherent demodulation of an FSK signal.

The probability of error, P_e, for noncoherently detected FSK is:

$$P_e = \frac{1}{2} e^{-\left(\frac{E_b}{N_o}\right)}$$

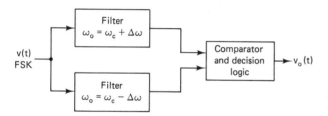

Figure 6-1 Non-coherent FSK demodulator.

In this expression, $N_o/2$ is the noise spectral density (AWGN), and E_b is the average bit energy for FSK, which is given by

$$E_b = \frac{A^2 T}{2}$$

where A is the peak carrier amplitude, and T is the bit period.

Example 6-1

A noncoherent FSK system is modulated by a PCM signal bit stream and operates at a frequency of 10 MHz. The frequency deviation is 850 Hz and the system data rate is 110 bits/sec. The peak-to-peak carrier amplitude is 2.0 volts and the double-sized noise spectral density is 1.0×10^{-4} volts^2/Hz. Calculate the bit error rate of this system.

Solution

$$A = \frac{V_{\text{P-P}}}{2} = \frac{2.0}{2} = 1.0 \text{ volts}$$

$$T = \frac{1}{f_b} = \frac{1}{110} = 9.09 \times 10^{-3} \text{ sec}$$

$$E_b = \frac{A^2 T}{2} = \frac{(1.0)(9.09 \times 10^{-3})}{2} = 4.55 \times 10^{-3} \text{ joules}$$

$$\frac{N_o}{2} = 1 \times 10^{-4} \text{ volts}^2/\text{Hz}$$

$$N_o = 2 \times 10^{-4} \text{ volts}^2/\text{Hz}$$

$$P_e = \text{BER} = \frac{1}{2} e^{-\left[\frac{4.55 \times 10^{-3}}{(2)(2 \times 10^{-4})}\right]} = \frac{1}{2} e^{-11.38}$$

$$= 5.74 \times 10^{-6}$$

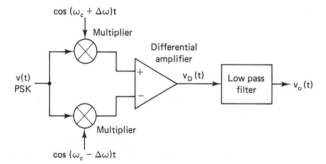

Figure 6-2 Synchronous FSK demodulator.

Figure 6-2 illustrates synchronous or coherent demodulation of an FSK signal. If the received FSK signal is

$$v(t) = A \cos (\omega_c + \Delta\omega)t$$

then the output of the differential amplifier is

$$v_D(t) = \frac{A}{2} - \frac{A}{2} [\cos 2\Delta\omega t + \cos 2\omega_c t - \cos 2(\omega_c + \Delta\omega)t]$$

If the received FSK signal is

$$v(t) = A \cos (\omega_c - \Delta\omega)t$$

then the output of the differential amplifier is

$$v_D(t) = -\frac{A}{2} + \frac{A}{2} [\cos 2\Delta\omega t + \cos 2\omega_c t - \cos 2(\omega_c - \Delta\omega)t]$$

The low-pass filter retains only the two quasi-DC terms, $A/2$ and $-A/2$, which represent the required binary output.

Synchronous demodulation is equivalent to a matched filter demodulator and the probability of error for this case is given by:

$$P_e = \frac{1}{2} \operatorname{erfc} \left(\sqrt{\frac{0.6 E_b}{N_o}} \right)$$

In this expression, E_b is the average bit energy, $N_o/2$ is the noise spectral density and erfc is the complementary error function.

Example 6-2

The FSK signal of Example 6-1 is applied to a synchronous demodulator. Calculate the improved bit error rate and compare it to the noncoherently demodulated case.

Solution

$$E_b = 4.55 \times 10^{-3} \text{ (from Example 6-1)}$$

$$N_o = 2 \times 10^{-4} \text{ (from Example 6-1)}$$

$$\text{BER} = P_e = \frac{1}{2} \text{ erfc} \left(\sqrt{\frac{(0.6)(4.55 \times 10^{-3})}{2 \times 10^{-4}}} \right) = \frac{1}{2} \text{ erfc } (3.69)$$

From Table 4-3, erfc $(3.69) \approx 1.9 \times 10^{-7}$

$$P_e = \frac{1.9 \times 10^{-7}}{2} = 9.5 \times 10^{-8} \text{ (coherent demodulation)}$$

Compared to

$$P_e = 5.7 \times 10^{-6} \text{ (noncoherent demodulation)}$$

Noncoherent frequency shift keying requires approximately a 2 dB increase in signal energy to equal the bit error rate performance of coherent FSK. The 2 dB penalty is somewhat compensated for by the simpler circuitry required for the noncoherent case.

6.2 BINARY PHASE SHIFT KEYING

Binary phase shift keying is a digital modulation method in which the phase of the carrier is set at 0° phase shift or 180° phase shift (or +90° and −90° relative to a carrier reference phase) depending upon the binary modulating signal. Binary phase shift keying (BPSK) is commonly just called *phase shift keying* (PSK). The PSK waveform may be represented as

$$v(t) = A \cos \left[\omega_c t + \phi(t) \right]$$

where A is a constant amplitude and $\phi(t)$ is either 0 or π radians ($\pm \pi/2$). A phase shift of π radians or 180° is equivalent to multiplication by −1, so a PSK signal may also be written as

$$v(t) = A \cos \omega_c t$$

$$v(t) = - A \cos \omega_c t$$

The two equations correspond to the two states in a binary modulation signal $m(t)$. If the binary modulating signal is designed so that its two states are represented

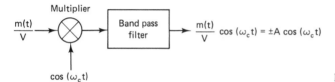

Figure 6-3 PSK signal generation.

by $m(t)$ taking on the values $+V$ and $-V$, the two equations may be written in a combined form:

$$v(t) = \frac{m(t)}{V} A \cos \omega_c t$$

This equation suggests a simple technique for generation of a PSK signal with a multiplier circuit. The ideal multiplier illustrated in Figure 6-3 provides an output signal that is the product of the two input signals.

A PSK signal must be demodulated synchronously. Figure 6-4 illustrates synchronous demodulation. Note that a phase shift θ has been indicated on the PSK signal in Figure 6-4. This phase shift is a function of the path length between transmitter and receiver. The output of the ideal synchronous demodulator is the product of its two input signals or

$$v_o(t) = \frac{1}{2} \frac{m(t)A}{V} + \frac{1}{2} \frac{m(t)A}{V} \cos 2 (\omega_c t + \theta)$$

A low-pass filter will recover the quasi-DC term, which is the desired modulating signal $m(t)$. Note that $m(t)$ will experience some distortion, since the filter will attenuate some of its higher frequency components. This distortion is not important, however, as it is necessary to know only if a logical 1 or logical 0 is present. If the low-pass filter is optimum in the sense that it maximizes the signal in the presence of noise, then it is a matched filter.

One input to the synchronous demodulator requires the following signal:

$$\cos (\omega_c t + \theta)$$

Since the phase shift θ is a function of transmitter-to-receiver distance, the signal must be derived at the receiver from the received PSK waveform. Figure 6-5 illustrates a simple technique for deriving this synchronizing signal.

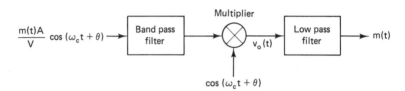

Figure 6-4 Synchronous demodulation of PSK signal.

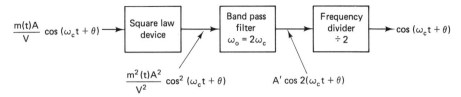

Figure 6-5 Synchronizing circuit for demodulation of PSK.

known as a correlation receiver. In the correlation receiver, the bit length is T.
All capacitors in the integrator are discharged at the beginning of each bit interval.
The correlation receiver can be shown to have equivalent performance to a matched
filter receiver.

The probability of error for a PSK system with a matched filter at the receiver
is

$$P_e = \frac{1}{2} \, \text{erfc} \left(\sqrt{\frac{E_b}{N_o}} \right)$$

In this expression, E_b is given by

$$E_b = \frac{A^2 T}{2}$$

Example 6-3

Calculate the probability of error for the system in Example 6-1 if phase shift keying
is used instead of FSK and compare it to the bit error rate for coherent and
noncoherent FSK.

Solution $E_b = 4.55 \times 10^{-3}$ (from Example 6-1)

$N_o = 2 \times 10^{-4}$ (from Example 6-1)

$$P_e = \frac{1}{2} \, \text{erfc} \left(\sqrt{\frac{4.55 \times 10^{-3}}{2 \times 10^{-4}}} \right) = \frac{1}{2} \, \text{erfc} \, (4.77)$$

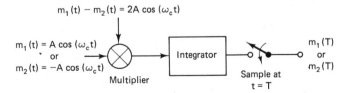

Figure 6-6 Correlation receiver.

From Table 4-3,

$$\text{erfc } (4.77) \approx 1.55 \times 10^{-11}$$

$$P_e = \frac{1.55 \times 10^{-11}}{2} = 7.75 \times 10^{-12} \text{ (PSK)}$$

compared to

$$P_e = 8 \times 10^{-8} \text{ (coherent FSK)}$$

$$P_e = 5.7 \times 10^{-6} \text{ (noncoherent FSK)}$$

Digital modulation methods are often compared on the basis of power efficiency and spectral efficiency. Power efficiency is defined in terms of the E_b/N_o required for an acceptable P_e performance. A common rule of thumb is that a digital modulation method is power efficient if an E_b/N_o of less than 14 dB is required to achieve a P_e of 10^{-8}. Spectral or bandwidth efficiency is normally expressed as the number of bits/second/Hertz (b/s/Hz). The rule of thumb for spectral efficiency is that a modulation technique is bandwidth efficient if it has an efficiency greater than 2 b/s/Hz. Binary phase shift keying is a power-efficient modulation method, because its P_e is approximately 7.5×10^{-13} for a 14 db E_b/N_o. BPSK is not spectrally efficient, however, as it theoretically requires 1 b/s/Hz for transmission.

Example 6-4

Compute the minimum theoretical bandwidth necessary for a 60 Mbit/sec data stream used in a transmitter with PSK modulation.

Solution

PSK transmission requires 1 b/s/Hz, so the system bandwidth would have to be 60 MHz.

6.3 QUADRATURE PHASE SHIFT KEYING

Quadrature phase shift keying (QPSK or 4PSK) is a modulation technique wherein the modulated signal has four distinct phase states as compared to BPSK (2PSK) where the signal has only two states. QPSK systems are more efficient spectrally than BSPK systems, as their theoretical spectral efficiency is 2 b/s/Hz.

The four phase states in QPSK are normally defined as $\pi/4$ (45°), 3 $\pi/4$ (135°), 5 $\pi/4$ (225°) and 7 $\pi/4$ (315°). In QPSK, the modulator takes consecutive *dibits* (pairs of bits) and maps them into one of the four phase states. The normal mapping code is the Gray code, which has the advantage that adjacent phase states

TABLE 6-1 GRAY CODE
ASSIGNMENTS IN QPSK

Dibit	Phase
11	45°
01	135°
00	225°
10	315°

differ by only one bit.   Table 6-1 illustrates a standard Gray code assignment in QPSK.

The phase states indicated in Table 6-1 are maintained for an interval T_s (symbol duration) which equals the duration of two bits ($T_s = 2T$).   The four Gray code states may be described mathematically as:

$$m_{11}(t) = A \cos (\omega_c t + \pi/4)$$

$$m_{01}(t) = A \cos (\omega_c t + 3\pi/4)$$

$$m_{00}(t) = A \cos (\omega_c t + 5\pi/4)$$

$$m_{10}(t) = A \cos (\omega_c t + 7\pi/4)$$

Figure 6-7 illustrates a QPSK transmitter block diagram.

In Figure 6-7, $I(t)$ is called the in-phase channel, while $Q(t)$ is called the quadrature channel.   (Quadrature indicates a phase shift of 90° or $\pi/2$ radians.) Figure 6-8 illustrates the relationship between the input data stream and the I and Q channel data streams.

Note that the symbol rate (the baud rate in this case) is equal to one-half of

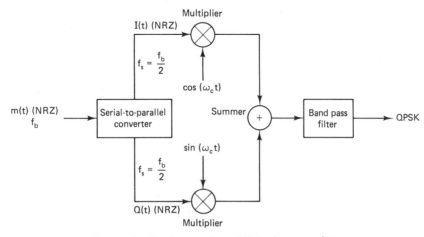

Figure 6-7   Quadrature phase shift keying transmitter.

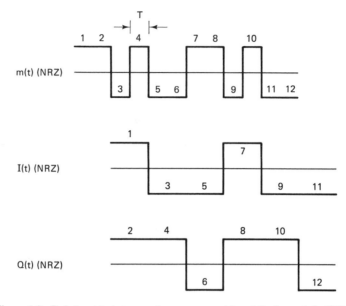

Figure 6-8 Relationship between data stream and I and Q channels in QPSK.

the input data stream rate or $T_s = 2T$. The numbers in Figure 6-8 indicate that bit 1 is delayed and expanded as symbol 1 on the I channel, bit 2 is delayed and expanded as symbol 2 on the Q channel and so on. The first dibit in Figure 6-8 is 11, so the data on the I and Q channels during the first symbol period is 11; the second dibit is 01 which appears on I and Q during the second symbol period; and so on. It is also important to note that all of the binary data streams in this system are nonreturn to zero (NRZ) which can only take on the normalized values of $+1$ and -1.

The I data stream and the Q data stream are applied to the inputs of separate multipliers acting as balanced modulators. The $I(t)$ signal is multiplied by a locally generated carrier (cos $\omega_c t$) and the $Q(t)$ signal is multiplied by the local carrier phase shifted by 90° or $\pi/2$ radians (sin $\omega_c t$). This arrangement is effectively two parallel BPSK systems. Figure 6-9 illustrates vector diagrams of the signals at the outputs of the two balanced modulators. Figure 6-9 illustrates that the output of

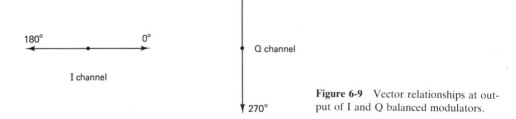

Figure 6-9 Vector relationships at output of I and Q balanced modulators.

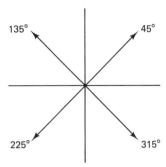

Figure 6-10   Vector sum of I and Q channels.

the I channel is a BPSK signal with phase states 0° or 180°, and the output of the Q channel is a BPSK signal with phase states 90° or 270°. Figure 6-10 illustrates the vector sum of these two channels; this is the input to the band pass filter at the output of the summer, which is used to limit the spectrum of the QPSK signal.

Figure 6-11 illustrates a QPSK receiver block diagram. The bandpass filter is used to reduce out-of-band noise and interference. The output of the filter is split into I and Q channels and demodulated coherently with a locally generated carrier (cos $\omega_c t$) recovered from the received signal for the I channel and demodulated coherently with the locally generated carrier shifted in phase by 90° (sin $\omega_c t$) for the Q channel. The low-pass filters remove the higher order products generated by the coherent demodulation process to produce the original baseband I and Q signals. The comparator outputs a logical 1 if the sampled value is positive; otherwise it provides a logical 0 output. The comparator is sampled at the symbol rate $(f_b/2)$. The symbol rate clock is synchronized to the I and Q channels. Finally the binary output data stream is produced by the parallel-to-serial converter.

The probability of a bit error in a Gray-coded QPSK system is:

$$P_e = \frac{1}{2} \operatorname{erfc}\left(\sqrt{\frac{E_b}{N_o}}\right)$$

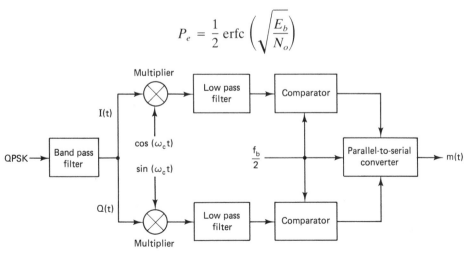

Figure 6-11   Quadrature phase shift keying receiver.

In this expression the bit energy is computed as follows:

$$E_b = A^2 T$$

The E_b/N_o requirement for a given bit error rate is nearly the same for BPSK and QPSK; however, the QPSK system only requires one-half of the RF bandwidth of the BPSK system.

In digital communications, error probabilities are normally computed as a function of the carrier-to-noise ratio at the receiver input. An important relationship between E_b/N_o and the carrier-to-noise ratio is given by

$$\frac{E_b}{N_o} = \frac{C}{N} \cdot \frac{BW}{f_b}$$

In this expression, C/N is the carrier-to-noise ratio, BW is the receiver noise bandwidth, and f_b is the bit rate. Bit error rate curves are often plotted in terms of C/N ratio rather than E_b/N_o.

When QPSK is compared to BPSK, the above relationship indicates that BPSK has a 3 dB advantage over QPSK in terms of the carrier-to-noise ratio required for a given P_e even though the E_b/N_o requirements are nearly the same. This advantage is illustrated by the following example.

Example 6-5

Compare the carrier-to-noise ratios required to implement a 120 Mbit/sec data communications system using BPSK and QPSK if a bit error rate of 10^{-7} is desired.

Solution

BPSK System:

$$P_e = \frac{1}{2} \operatorname{erfc} \left(\sqrt{\frac{E_b}{N_o}} \right)$$

$$10^{-7} = \frac{1}{2} \operatorname{erfc} \left(\sqrt{\frac{E_b}{N_o}} \right)$$

From Table 4-3,

$$\frac{E_b}{N_o} \approx 13.5$$

$$f_b = 120 \text{ Mbit/sec}$$

$$BW = 120 \text{ MHz}$$

QPSK System:

$$P_e = \frac{1}{2} \, \text{erfc} \left(\sqrt{\frac{E_b}{N_o}} \right)$$

$$10^{-7} = \frac{1}{2} \, \text{erfc} \left(\sqrt{\frac{E_b}{N_o}} \right)$$

$$\frac{E_b}{N_o} \approx 13.5 \text{ (same as for BPSK)}$$

$$f_b = 120 \text{ Mbit/sec}$$

$$BW = 60 \text{ MHz}$$

Using the relationship between E_b/N_o and C/N:

$$\frac{C}{N} = \left[\frac{f_b}{BW} \right] \left[\frac{E_b}{N_o} \right]$$

$$\text{BPSK:} \frac{C}{N} = \left[\frac{120}{120} \right] [13.5] = 13.5 = 11.3 \text{ dB}$$

$$\text{QPSK:} \frac{C}{N} = \left[\frac{120}{60} \right] [13.5] = 27.0 = 14.3 \text{ dB}$$

Thus the QPSK system requires a 3 dB greater *C/N* ratio for equivalent bit error rate performance.

6.4 *M*-ARY PHASE SHIFT KEYING

M-ary phase shift keying is a general term used to refer to phase shift keying modulation techniques. In normal practice 2PSK is referred to as BPSK or just PSK, 4PSK is referred to as QPSK, and phase shift keying with a higher number of phase states is often lumped under the term *M*-ary PSK or MPSK. The *M*-ary systems that are used in practice include 8PSK, 16PSK, and sometimes 32PSK. The higher order systems are increasingly more spectrally efficient, require more complex modem circuitry and require a higher carrier-to-noise ratio for a given bit error rate. Table 6-2 summarizes the theoretical spectral efficiency of the various *M*-ary PSK systems used in practice.

The theoretical bandwidths listed in Table 6-2 can be approached in practical systems but cannot be equaled because the required filters cannot be designed with infinitely sharp rolloffs.

TABLE 6-2 THEORETICAL
BANDWIDTH REQUIREMENTS
FOR *M*-ARY PSK

Modulation	Theoretical Bandwidth
BPSK	1 b/s/Hz
QPSK	2 b/s/Hz
8PSK	3 b/s/Hz
16PSK	4 b/s/Hz
32PSK	5 b/s/Hz

TABLE 6-3 APPROXIMATE CARRIER-TO-NOISE
RATIO REQUIRED FOR $P_e = 10^{-7}$
FOR *M*-ARY PSK

Modulation	Approximate C/N Ratio ($P_e = 10^{-7}$)
BPSK	11.5 dB
QPSK	14.5 dB
8PSK	19.5 dB
16PSK	25.5 dB
32PSK	32.5 dB

Table 6-3 summarizes the approximate carrier-to-noise ratio in dB required for a bit error rate of 10^{-7} for *M*-ary PSK systems.

Higher order PSK systems are becoming increasingly important in applications, such as satellite communications, where the spectrum is crowded and the signal bandwidth must be minimized. The increased signal-to-noise ratios required for these higher order systems are becoming easier to achieve with the continuing improvement in low-noise receiving amplifiers and high-power transmitting amplifiers.

TABLE 6-4 PERFORMANCE OF *M*-ARY PSK SYSTEMS COMPARED TO
SHANNON-HARTLEY LIMIT

Modulation	Approximate C/N Ratio ($P_e = 10^{-7}$)	Theoretical Bandwidth	Approximate Shannon-Hartley Bandwidth
BPSK	11.5 dB	1 b/s/Hz	3.9 b/s/Hz
QPSK	14.5 dB	2 b/s/Hz	4.8 b/s/Hz
8PSK	19.5 dB	3 b/s/Hz	6.3 b/s/Hz
16PSK	25.5 dB	4 b/s/Hz	7.6 b/s/Hz
32PSK	32.5 dB	5 b/s/Hz	9.9 b/s/Hz

It is of interest to compare the spectral efficiency of *M*-ary PSK systems with the absolute limit calculated from the Shannon-Hartley law discussed in Chapter 2. Table 6-4 provides a summary of the performance of *M*-ary PSK systems compared to the Shannon-Hartley limit for a $P_e = 10^{-7}$. From this table it can be seen that the BPSK signal requires about 3.9 times the Shannon-Hartley predicted bandwidth, while the 32PSK signal only requires 1.98 the bandwidth predicted by Shannon-Hartley.

6.5 DIFFERENTIAL BINARY PHASE SHIFT KEYING

Differential binary phase shift keying (DBPSK or DPSK) is a digital modulation technique that avoids the necessity of providing a synchronous locally generated carrier at the receiver for demodulation. DPSK systems generate a differential binary data stream that provides information on changes in the original information data stream. Figure 6-12 provides a block diagram of a DPSK transmitter. In Figure 6-12, *m*(*t*) is the information data stream, and *d*(*t*) is the differential data stream. The level shift circuit converts *d*(*t*) to an NRZ data stream with normalized pulse amplitudes of ±1. The multiplier and bandpass filter are identical in function to those elements in a BPSK transmitter. Figure 6-13 illustrates the operation of the logic and delay circuits.

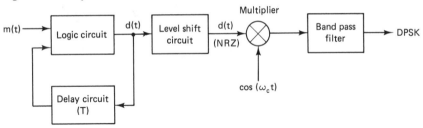

Figure 6-12 Differential binary phase shift keying transmitter.

The logic circuit compares information bits *m*(*t*) and *d*(*t* − *T*) and provides a logical 1 output if the two bits are identical and a logical 0 output if they are different. Note that an arbitrary bit has been indicated on the *d*(*t*) waveform. This bit may be either logical 1 or logical 0 and is used as a reference bit for the first comparison. In practice, *d*(*t*) may be held at a constant voltage level corresponding to logical 1 or logical 0 until an *m*(*t*) signal is present. The logic equation describing this operation is $d(t) = \overline{m(t) \oplus d(t - T)}$, where ⊕ indicates the exclusive-or operation.

Figure 6-14 illustrates a technique for demodulation of DPSK. In Figure 6-14, the received DPSK signal is multiplied by itself delayed by *T*, the duration of a bit. This technique results in a signal at the output of the multiplier that contains the product of the received DPSK signal and its delayed version as well

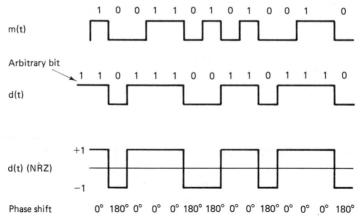

Figure 6-13 Data stream relationships in DPSK.

as a term at $2\omega_c$ that is removed by the low-pass filter. Figure 6-15 illustrates recovery of the original bit stream $m(t)$. The illustrated bit streams are consistent with Figure 6-13. Note that the product waveform in Figure 6-15 is the original information signal $m(t)$ in NRZ form.

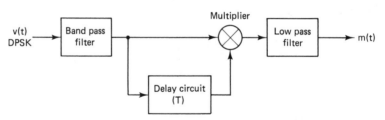

Figure 6-14 Differential binary phase shift keying receiver.

DPSK does not perform as well as PSK with respect to bit error rate. The probability of error for a DPSK signal is given by

$$P_e = \frac{1}{2}e^{-\left(\frac{E_b}{N_o}\right)}$$

In this expression E_b is given by

$$E_b = \frac{A^2T}{2}$$

Example 6-6

A 4 GHz, 90 Mbit/sec communications link is implemented with differential phase shift keying modem. If the bit sequence given in Figure 6-16 is transmitted over the system, sketch the differential waveform and calculate the bit error rate for this link. Assume $A = 10$ volts and $N_o = 4 \times 10^{-8}$ volts²/Hz.

Solution

The differential waveform is illustrated in Figure 6-16.

$$T = \frac{1}{f_b} = \frac{1}{90 \times 10^6} = 1.11 \times 10^{-8} \text{ sec}$$

$$E_b = \frac{A^2T}{2} = \frac{(10)^2(1.11 \times 10^{-8})}{2} = 5.56 \times 10^{-7} \text{ joules}$$

$$P_e = \frac{1}{2} e^{-\left(\frac{E_b}{N_o}\right)} = \frac{1}{2} e^{-\left(\frac{5.56 \times 10^{-7}}{4 \times 10^{-8}}\right)} = 4.6 \times 10^{-7}$$

1 1 0 1 1 1 0 0 1 1 0 1 1 1 0
+1
d(t) (NRZ)
−1
1 1 0 1 1 1 0 0 1 1 0 1 1 1 0
+1
d(t − T) (NRZ)
−1
1 0 0 1 1 0 1 0 1 0 0 1 1 0
+1
d(t) d(t − T) (NRZ)
−1

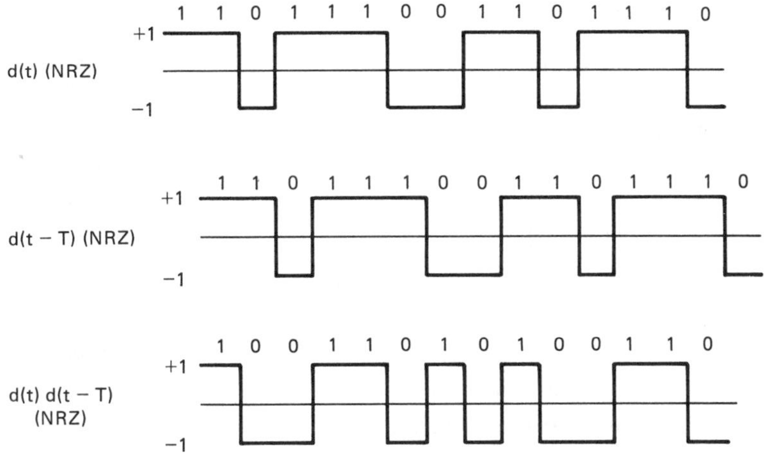

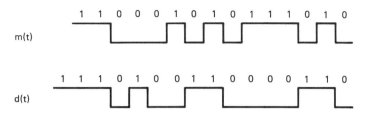

Figure 6-15 DPSK demodulation waveforms.

Bit sequence: 1 1 0 0 0 1 0 1 0 1 1 1 0 1 0
1 1 0 0 0 1 0 1 0 1 1 1 0 1 0
m(t)
1 1 1 0 1 0 0 1 1 0 0 0 0 1 1 0
d(t)

Figure 6-16 Bit sequence for Example 6-6.

6.6 MINIMUM SHIFT KEYING

Minimum shift keying (MSK) is a digital modulation method closely related to frequency shift keying. MSK has a spectral efficiency of 2 b/s/Hz like QPSK and exhibits the same bit error rate of

$$P_e = \frac{1}{2} \, \mathrm{erfc} \left(\sqrt{\frac{E_b}{N_o}} \right)$$

In this expression E_b is given by

$$E_b = \frac{A^2 T}{2}$$

MSK has the advantage (for some applications) that the modulated signal has a constant carrier amplitude, which means it can be limited in the same fashion as analog FM signals.   An additional advantage of MSK is that its spectrum falls off faster than QPSK.   Figure 6-17 illustrates an MSK transmitter block diagram. The serial-to-parallel converter distributes alternate bits in the $m(t)$ data stream to the I and Q channels with a delay of T in the Q channel.   The symbol duration T_s in the I and Q channels is $2T$.

The sinusoidal pulse-shaping circuits convert the rectangular bit streams in the I and Q channels to half sinusoidal pulses which are filtered, applied to the multipliers and summed.   Figure 6-18 illustrates the relationships between pulses in the input data stream and the I and Q channels.   The I channel pulse shaper output may be represented mathematically as

$$\cos \left(\pm \frac{\pi}{2T} t \right)$$

The Q channel pulse shaper output is

$$\cos \left[\pm \frac{\pi}{2T}(t - T) \right] = \sin \left(\pm \frac{\pi}{2T} t \right)$$

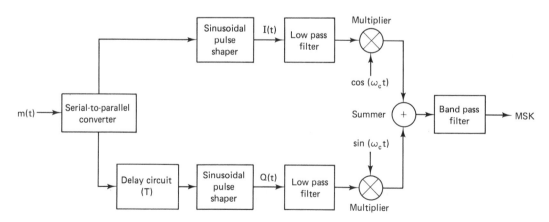

Figure 6-17   Minimum shift keying transmitter.

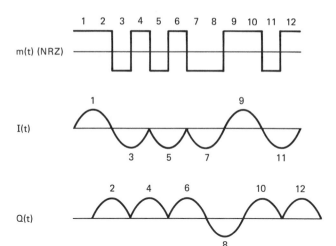

Figure 6-18 Pulse relationships in MSK transmitter.

As discussed in Section 6.1, an FSK signal may be represented mathematically as:

$$v(t)_{\text{FSK}} = A \cos (\omega_c \pm \Delta\omega)t$$

This expression may be expanded using a trigonometric identity to:

$$v(t)_{\text{FSK}} = A \cos (\omega_c \pm \Delta\omega)t$$

$$= A \cos (\pm\Delta\omega t) \cos (\omega_c t) - A \sin (\pm\Delta\omega t) \sin (\omega_c t)$$

For the special case of MSK, the following relationship between deviation and bit duration must hold:

$$\Delta\omega = \frac{\pi}{2T}$$

The mathematical representation of an MSK signal then becomes:

$$v(t)_{\text{MSK}} = A \cos \left(\pm\frac{\pi}{2T}t \right) \cos (\omega_c t) - A \sin \left(\pm\frac{\pi}{2T}t \right) \sin (\omega_c t)$$

The transmitter in Figure 6-17 implements this expression.

Figure 6-19 illustrates an MSK receiver block diagram. In the MSK receiver, the $I(t)$ and $Q(t)$ baseband signals are recovered by synchronous demodulation and filtering. The sinusoidal pulses are squared up by the comparators, the I channel bit stream is delayed by T to compensate for the Q channel delay in the transmitter, and both are combined into the original information data stream by the parallel-to-serial converter.

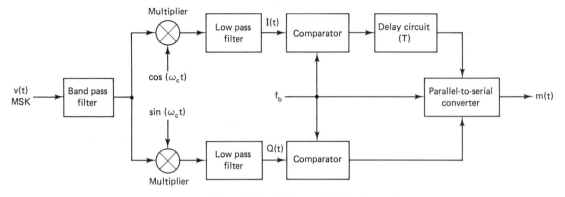

Figure 6-19   Minimum shift keying receiver.

Example 6-7

An S-band, 60 Mbit/sec, MSK data link is used to transmit computer data.   Calculate the deviation in MHz for this system.

Solution

$$T = \frac{1}{f_b} = \frac{1}{6 \times 10^7} = 1.67 \times 10^{-8}$$

$$\Delta\omega = \frac{\pi}{2T} = \frac{3.14}{(2)(1.67 \times 10^{-8})}$$

$$= 94 \times 10^6 \text{ radians/sec} = 15 \text{ MHz}$$

6.7 QUADRATURE AMPLITUDE MODULATION

Quadrature amplitude modulation (QAM), sometimes called *amplitude phase keying* (APK) or *multiple amplitude phase keying* (MAPK), is a form of digital modulation that normally operates with more than two states.   In QAM, the information is contained in both the amplitude and the phase of the modulated signal.   A popular implementation of QAM is 16QAM, which has a theoretical spectral efficiency of 4 b/s/Hz.   16QAM has 16 different signal states.   Figure 6-20 illustrates a 16QAM transmitter block diagram, and figure 6-21 illustrates a 16QAM receiver block diagram.

The bit error rate performance of 16QAM lies between that of 8PSK and 16PSK.   For a $P_e = 10^{-7}$, 8PSK requires a carrier-to-noise ratio of about 19.5 dB, while 16PSK requires a C/N of approximately 23.5 dB.   16QAM requires a C/N of 22 dB for the same 10^{-7} bit error rate performance.   Since 16PSK and 16QAM exhibit the same theoretical spectral efficiency of 4 b/s/Hz, and 16QAM

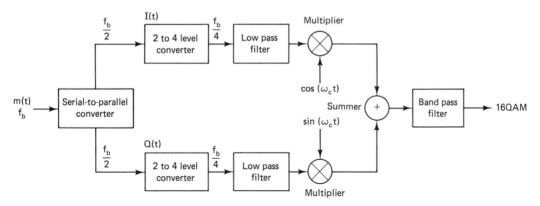

Figure 6-20   16 quadrature amplitude modulation transmitter.

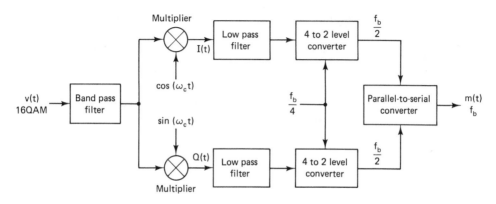

Figure 6-21   16 quadrature amplitude modulation receiver.

requires a lower *C/N* ratio, 16QAM is a more effective modulation method than 16PSK.

EXERCISES

1. An FSK signal is described by the equation $v(t) = 10 \cos (4.77 \times 10^6 \pm 8.5 \times 10^2)t$. Calculate the center frequencies of the two bandpass filters required to demodulate this signal noncoherently.
2. Data at 300 bps (bits/sec) is to be transmitted over an 8.5 MHz, noncoherently demodulated FSK system with a deviation of 170 Hz. Calculate the bit error rate if the carrier amplitude is 1.5 volts and the noise spectral density is 1.0×10^{-4} volts²/Hz.
3. Data at 110 bps is to be transmitted over a 12 MHz coherently demodulated

FSK system with a deviation of 850 Hz. Calculate the bit error rate if the peak-to-peak carrier amplitude is 2.0 volts and the noise spectral density is 1.0 $\times$ 10^{-4} volts^2/Hz.

4. An S-band PSK system is used to transmit data at 1200 bps. If the carrier amplitude is 2.0 volts and the noise spectral density is 1.0 $\times$ 10^{-4} volts^2/Hz, calculate the probability of error and the bandwidth required for this system.

5. Compute the carrier-to-noise ratio required for a 60 Mbit/sec, 4 GHz, QPSK data communications system if a bit error rate of 10^{-8} is desired.

6. Compute the carrier-to-noise ratio required for an 80 Mbit/sec, 6 GHz, PSK data communications system if a bit error rate of 10^{-7} is required.

7. The bit sequence 11001110001 (LSB sent first) is transmitted over a 14 GHz DPSK system. Sketch the differential waveform for this system.

8. A 90 Mbit/sec data stream is to be transmitted over a 2 GHz DPSK system. If the carrier amplitude is 8.0 volts and the noise spectral density is 3.6 $\times$ 10^{-8} volts^2/Hz, calculate the bit error rate.

9. Calculate the bandwidth required to transmit a 60 Mbit/sec data stream over a 4 GHz MSK data communications system.

10. Calculate the deviation in MHz for the MSK system in Exercise 9.

11. Compute the bandwidth necessary for a 90 Mbit/sec, 14 GHz, 8PSK data communications system.

12. Compute the bandwidth required if the 90 Mbit/sec data stream is transmitted over a 14 GHz 16QAM system.

7

Spread Spectrum
Methods

Spread spectrum is a technique in which a transmitted signal is spread over a frequency range that is much greater than the minimum bandwidth required for information transmission. The purpose of a spread spectrum system is to improve the bit error rate in the presence of excessive noise or interference. Spread spectrum is based upon the Shannon-Hartley law, introduced in Chapter 2, which illustrates that signal-to-noise ratio may be traded off against system bandwidth. A number of techniques exist to implement spread spectrum including direct sequence, frequency hopping, chirp, and time hopping. Some systems utilize a combination of techniques giving rise to such hybrid systems as time–frequency hopping. Spread spectrum systems exhibit what is known as *process gain*, the ratio of the signal-to-noise ratio at the output of a spread spectrum modulator to the signal-to-noise ratio at its input. The process gain in a spread spectrum system is given approximately by

$$G_p = 10 \log \frac{BW_{\mathrm{RF}}}{f_b}$$

In this expression, G_p is the process gain in dB, BW_{RF} is the RF bandwidth after spreading and f_b is the information-signal data rate. Spread spectrum process gain is analogous to the FM improvement factor in an analog wideband FM system. Another quantity of interest in spread spectrum systems is the ability of the system to operate in the presence of interference, called the jamming margin. Jamming margin is given by

$$M_j = G_p - L_s + (S/N)_{\mathrm{min}}$$

where M_j is the jamming margin in dB, G_p is the process gain, L_s is the system loss in dB, and $(S/N)_{\mathrm{min}}$ is the minimum required signal-to-noise ratio of the baseband digital signal in dB at the information output of the receiver.

Applications for spread spectrum systems in addition to signal-to-noise ratio and interference rejection improvements include the ability to selectively address specific receivers, the ability to "code division multiplex" several signals, the ability to hide signals as a result of the low-power spectral density that occurs (useful in military applications), and the ability to prevent eavesdropping by casual listeners.

7.1 DIRECT SEQUENCE SYSTEMS

The most widely used technique for generating a spread spectrum signal is the direct sequence method. In this method, the carrier is modulated by a digital code sequence whose bit rate is much higher than the information-signal bit rate. Most spread spectrum systems utilize BPSK or QPSK modulation, although MSK is popular when a fixed code rate is required and minimum bandwidth is necessary. Since each of these modulation methods utilizes balanced modulators, the carrier is suppressed, which aids in signal hiding and allocates available power for information transmission rather than for an unnecessary carrier.

The baseband information may be embedded in the spread spectrum signal either by modulating the carrier before spreading it or by adding the digitized information data stream to the spectrum-spreading code before its use as spreading modulation. Direct modulation of the carrier normally utilizes FM or PM, if the baseband signal is in analog form, or PSK, if it is in a digital format.

Example 7-1

A 10 GHz direct sequence spread spectrum system is used to transmit a 19.6 Kbaud data stream. The spreading code has a data rate of 10 Mbit/sec and the modulation method is BPSK. Calculate the process gain of this system.

Solution

BPSK requires 1 b/s/Hz; 10 Mb/sec requires 10 MHz bandwidth.

$$G_p = 10 \log \frac{10 \times 10^6}{19.6 \times 10^3} = 10 \log 510 = 27.1 \text{ dB}$$

Example 7-2

The modulation method in Example 7-1 is changed to QPSK. Calculate the resulting process gain with this new system.

Solution

QPSK requires 2 b/s/Hz; 10 Mb/sec requires 5 MHz bandwidth.

$$G_p = 10 \log \frac{5 \times 10^6}{19.6 \times 10^3} = 24.1 \text{ dB}$$

Example 7-3

If the system in Example 7-1 has system losses of 3 dB and requires a minimum signal-to-noise ratio of 10 dB, calculate the jamming margin.

Solution

$$M_j = G_p - L_s - (S/N)_{min}$$
$$M_j = 27.1 - 3 - 10 = 14.1 \text{ dB}$$

A spread spectrum receiver is implemented by correlating a locally generated version of the spreading code with the received signal. This process *despreads* the received signal and recovers the original baseband information. Any incoming signals that are not synchronized with the receiver's coded reference will be spread themselves and are easily removed by a post-correlation bandpass filter. Undesired

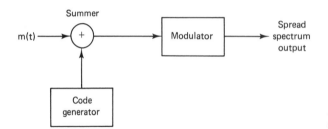

Figure 7-1 Direct sequence spread spectrum transmitter.

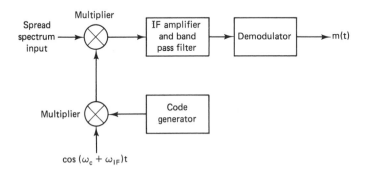

Figure 7-2 Direct sequence spread spectrum receiver.

signals appear as a slight increase in noise. Transmitted spread spectrum signals also appear as noise when received on a conventional or nonsynchronous receiver and thus cause minimal interference to other signals using the same portion of the frequency spectrum.

Figure 7-1 illustrates a simplified direct sequence spread spectrum transmitter block diagram and Figure 7-2 illustrates the corresponding simplified receiver. In Figure 7-1, the digitized information signal $m(t)$ is added to the signal produced by the code generator and the composite data stream modulates a carrier using a digital modulation method such as BPSK, QPSK or MSK.

In Figure 7-2, the code generator generates the same code as the transmitter, which is multiplied with a locally generated carrier offset by the intermediate frequency (IF) and mixed with the input signal to despread it. The despread signal is demodulated with a BPSK, QPSK or MSK demodulator depending upon the modulation technique used at the transmitter.

7.2 FREQUENCY HOPPING SYSTEMS

A frequency hopping spread spectrum system is an FSK system with many frequencies (often thousands) rather than just two. It is implemented with a digital frequency synthesizer whose output is determined by a code generator. At the

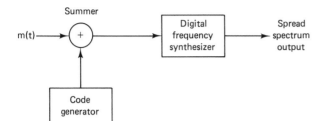

Figure 7-3 Frequency-hopping spread spectrum transmitter.

receiver, the received spread spectrum signal is mixed (multiplied) with a locally generated hopped carrier that is controlled by the same code sequence as the transmitted signal but whose output frequency is offset by the desired intermediate frequency (IF).

Figure 7-3 illustrates a simplified block diagram of a frequency hopping spread spectrum transmitter. In Figure 7-3, the information data stream is added to the spreading code just as in a direct sequence system. The composite code is used to control a digital frequency synthesizer, a device that provides discrete frequencies in response to a digital input. Figure 7-4 illustrates a simplified block diagram of a frequency hopping spread spectrum receiver. In Figure 7-4, the code generator and digital frequency synthesizer produce a synchronous hopped carrier whose frequency is offset by the IF and mixed with the incoming signal to despread it. The despread signal is passed through an IF amplifier/bandpass filter and demodulated.

For a frequency hopping system whose discrete frequencies or channels are contiguous, the process gain is the same as indicated earlier or

$$G_p = 10 \log \frac{BW_{\mathrm{RF}}}{f_b}$$

In some frequency hopping systems, the channels are not contiguous and in this case, the process gain may be approximated by

$$G_p = 10 \log N$$

where N is the number of discrete frequencies or channels.

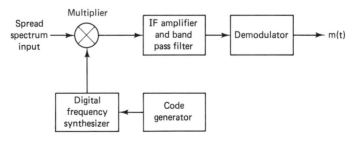

Figure 7-4 Frequency-hopping spread spectrum receiver.

Example 7-4

A frequency hopping spread spectrum is used to transmit a 4 Mbit/sec data stream. If the system utilizes 2000 non-contiguous channels, estimate the process gain.

Solution

$$G_p = 10 \log N = 10 \log (2000) = 33 \text{ dB}$$

The bit error rate resulting from interference in a frequency hopping system may be estimated from

$$P_e = \frac{J}{N}$$

where J is the number of jamming signals with power greater than or equal to the signal power and N is the number of channels.

Example 7-5

Estimate the bit error rate due to interference in Example 7-4 if 30 jamming signals are present.

Solution

$$P_e = \frac{J}{N} = \frac{30}{2000} = 1.5 \times 10^{-2}$$

This is generally an unacceptable error rate.

Bit error rate may be improved by increasing N or by sending more than one frequency per bit. Typical systems employ an odd number of frequencies per bit, such as 3 or 5, and base the received bit decision on a simple majority basis. Performance of such a system is greatly enhanced as compared to single frequency per bit systems. A redundant system such as the above would require an increase in the hop rate (by the number of frequencies per bit) as compared to the non-redundant system. The hop rate for a non-redundant system must be at least equal to the information data rate. The bandwidth of a dehopped signal is approximately equal to twice the information data rate.

Example 7-6

A 1200 baud data stream is to be sent over a non-redundant frequency hopping system. The maximum bandwidth for the spread spectrum signal is 10 MHz.

Calculate the maximum number of channels possible if no channel overlap occurs, and estimate the process gain for the system.

Solution

$$\text{Dehopped bandwidth} = 2 \times 1200 \text{ b/s} = 2.4 \text{ KHz}$$

$$\text{Number of channels} = \frac{10 \times 10^6}{2.4 \times 10^3} = 4167$$

$$G_p = 10 \log \frac{BW_{RF}}{f_b} = 10 \log \frac{10 \times 10^6}{1.2 \times 10^3} = 39.2 \text{ dB}$$

7.3 TIME HOPPING SYSTEMS

Time hopping is a spread spectrum technique that is rarely used by itself, but which is often used in conjunction with other spread spectrum systems such as direct sequence or frequency hopping. In time hopping, a code sequence is used to key a transmitter on and off. The information data stream is normally added to the code sequence prior to modulation, as is the case in direct sequence and frequency hopping systems. The primary advantage of time hopping systems is the reduced transmitter duty cycle. Simple time hopping systems are vulnerable to interference, since a continuous carrier at the transmitting frequency can block communications.

Time–frequency hopping spread spectrum systems are hybrid systems, most useful for applications where a large number of users with widely varying distances or powers must operate over a single link.

Time hopping–direct sequence systems are hybrid systems that are most often utilized when multiplexing is desired and when such multiplexing cannot be adequately implemented with code division multiplexing. Time hopping–direct sequence systems allow time division multiplexing of a number of data streams over a single link.

7.4 CHIRP SYSTEMS

Chirp is a form of spread spectrum modulation in which the frequency of a pulse is varied over its duration to increase its bandwidth. Coding is not normally used in chirp systems and the most often used sweep is linear. A chirp receiver requires a matched filter, which may be thought of as storing the transmitted signals and reassembling them to produce a stronger signal.

Chirp has found primary application in radar, but it can also be used in communications systems. Since a chirp system requires more than the minimum bandwidth required for transmission of the information data stream, it exhibits a process gain like that of the other spread spectrum implementations.

The two parameters of primary interest in chirp systems are the frequency sweep Δf and the time ΔT used to sweep over Δf. A key quantity derived from these two parameters is the compression ratio D, given by $D = \Delta f \Delta T$. A matched filter improves the voltage signal-to-noise ratio in a chirp system by a factor of $\sqrt{D}$.

7.5 CODE GENERATION AND SYNCHRONIZATION

Coding is the very heart of spread spectrum systems. The most suitable code for a spread spectrum system is a pseudorandom code sequence which is noise-like in its behavior but which is also deterministic. The only codes to be considered here are those used for spectrum spreading. Other codes, such as error control codes, are normally utilized in the baseband data stream.

The codes most utilized in spread spectrum systems are maximal linear codes, which may be generated in hardware by a shift register with feedback via logic gates or in software for microprocessor-based systems. An n-stage shift register can maximally provide $2^n - 1$ code bits. In a maximal code the number of ones and the numbers of zeros differ by only one. When NRZ waveforms are used, this means the DC level is negligible. This characteristic also ensures maximum carrier suppression in modulation techniques such as BPSK.

Figure 7-5 illustrates a simple code generator implemented with a three-stage shift register and feedback logic. The code lengths required for spread spectrum systems are much longer than the one illustrated in Figure 7-5, but increased code length requires only a longer shift register and additional feedback logic. The length of time to elapse before the code repeats itself may be as short as a few milliseconds or as long as several days for the "secure" codes in secret data transmissions. Feedback connections for a number of common maximal codes have been tabulated in publications on coding. Software generation of codes in microprocessor-based systems is almost trivial, but these systems are limited in speed and cannot be used when spreading is desired over a wide bandwidth.

Code synchronization between transmitter and receiver is a very complex problem that must be solved if the received signal is to be despread. In general, there are two types of synchronization problems in spread spectrum systems, code-phase synchronization and carrier-frequency synchronization. Even with stable frequency sources in the transmitter and the receiver, relative motion (Doppler shift) or propagation path length changes between transmitter and receiver can produce phase variations. Numerous techniques exist to initiate and maintain synchronization. A useful technique is to send a specific code sequence known as a preamble at the start of each transmission. The receiver searches for the preamble and "locks on" when it is found. Preamble codes range from a few hundred to a few thousand bits.

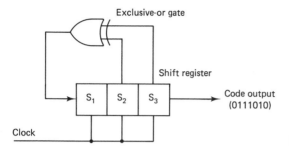

Figure 7-5 Simple code generator.

EXERCISES

1. A 9600 baud computer data link is implemented with an 18 GHz QPSK spread spectrum system with 500 MHz bandwidth. Calculate the process gain of this system.

2. Calculate the jamming margin of a 10 GHz spread spectrum system with a process gain of 30 dB, system losses of 5 dB and a minimum required signal-to-noise ratio of 6 dB.

3. Estimate the process gain of a 10 GHz frequency hopping BPSK spread spectrum communications system if the system utilizes 1800 non-contiguous channels.

4. A 4 GHz frequency hopping QPSK spread spectrum communications system with 2400 non-contiguous channels is subjected to 60 jamming signals with power greater than the signal power. Calculate the probability of error due to these jammers.

5. A spread spectrum system utilizing chirp modulation sweeps 10 MHz in 20 ms. Calculate the improvement in the signal-to-noise ratio for this system if a matched filter is used at the receiver.

8

Filters and Impedance Transformation Circuits

One of the most important types of electronic circuits utilized in communications systems is the filter, which may be classified as low pass, high pass, band pass, or band reject as discussed in Section 2.8. Filters may be implemented using passive elements such as resistors, capacitors and inductors or with an active element, such as an operational amplifier in conjunction with passive elements. Filters may also use digital techniques; examples are switched capacitor filters and traditional digital filters that use the computational power of the microprocessor.

In communications it is often necessary to transform a circuit from one impedance level to another. Impedance transformation circuits provide this transformation and in addition often provide some degree of filtering.

This chapter will only provide a brief introduction to filter and impedance transformation analysis and synthesis. Entire texts are available that discuss in detail the topics introduced here. Passive and active filter design tables and graphs are available in a number of handbooks.

8.1 FILTER TYPES

The most often used types are the Butterworth, Chebyshev, inverse Chebyshev, and elliptic. These types may be used to implement low pass, high pass, band pass, and band reject filters; the usual course is to synthesize a low pass filter and then to transform it into a high pass, band pass, or band reject filter, as necessary. Because the low pass filter is the basis for other filter characteristics, it will be used to describe the characteristics of the four major types.

The Butterworth low pass filter has an amplitude response given by

$$H(\omega) = \frac{A}{\sqrt{1 + \left(\dfrac{\omega}{\omega_c}\right)^{2n}}}$$

In this expression $H(\omega)$ is the amplitude response as a function of frequency, A is a constant, ω_c is the cutoff frequency, and n is the order of the filter. Figure 8-1 illustrates a normalized ($A = 1$) Butterworth low pass filter response for $n = 2$ and $n = 6$ as compared to an ideal low pass filter. Note that as the order of the filter increases, the response curve more nearly approaches the ideal case and that the response is flat near $\omega = 0$; for this reason, Butterworth filters are often called maximally flat filters.

The Chebyshev low pass filter has an amplitude response given by:

$$H(\omega) = \frac{A}{\sqrt{1 + \varepsilon^2 C_n^2\left(\dfrac{\omega}{\omega_c}\right)}}$$

In this expression $H(\omega)$ is the amplitude response as a function of frequency, A

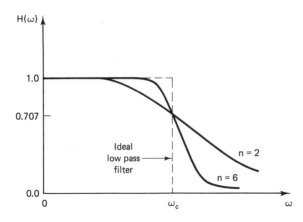

Figure 8-1 Normalized Butterworth low pass filter response.

and ε are constants, ω_c is the cutoff frequency, and $C_n\left(\dfrac{\omega}{\omega_c}\right)$ is the Chebyshev polynomial of degree n given by:

$$C_n(x) = \cos{(n \arccos{x})}$$

The resulting Chebyshev filter is of order n. Figure 8-2 illustrates a normalized ($A = 1$) Chebyshev low pass filter response for $n = 2$ and $n = 6$. In Figure 8-2, note that the higher order filter has a sharper cutoff, but that the number of ripples in the passband is greater. Note also that all ripples are equal in magnitude, giving rise to the often-used name, *equiripple filter*. The ripple width for a normalized Chebyshev filter is given by

$$RW = 1 - \frac{1}{\sqrt{1 + \varepsilon^2}}$$

Ripple width may be reduced by making ε smaller. Chebyshev filters are also characterized by the minimum allowable passband loss, given by

$$\alpha = 10 \log_{10}{(1 + \varepsilon^2)}$$

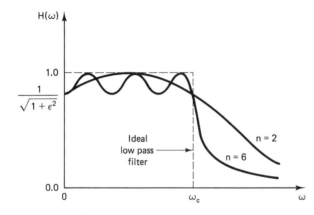

Figure 8-2 Normalized Chebyshev low pass filter response.

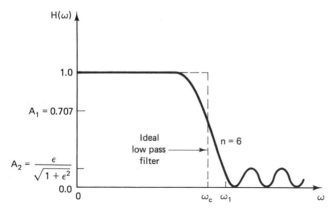

Figure 8-3   Normalized inverse Chebyshev low pass filter response.

In this expression α is the minimum allowable passband loss in dB.   A filter with $\alpha = 3$ is called a 3 dB Chebyshev filter.

The inverse Chebyshev filter has an amplitude response given by:

$$H(\omega) = \frac{\varepsilon\, C_n\!\left(\dfrac{\omega_1}{\omega}\right)}{1 + \varepsilon^2\, C_n^2\!\left(\dfrac{\omega_1}{\omega}\right)}$$

In this expression, $H(\omega)$ is the amplitude response as a function of frequency, ε is a constant, $C_n\!\left(\dfrac{\omega_1}{\omega}\right)$ is the Chebyshev polynomial of degree n, and ω_1 is the lower frequency limit of the stop band.   Figure 8-3 illustrates a normalized inverse Chebyshev low pass filter response for $n = 6$.   In Figure 8-3 note that the ripple has been translated to the stopband.

The elliptic filter's amplitude response is given in terms of the Jacobi elliptic function, which will not be discussed here.   Figure 8-4, however, graphically il-

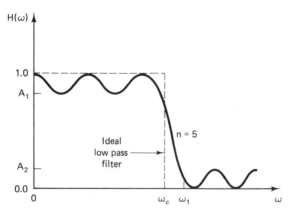

Figure 8-4   Normalized elliptic low pass filter response.

lustrates the response of a fifth order elliptic low pass filter. The elliptic filter has ripple in both the pass band and the stop band and exhibits a transition width ($\omega_1 - \omega_c$) which is superior to the other filter types for a given order.

The Butterworth, Chebyshev, inverse Chebyshev and elliptic low pass filters may be transformed into high pass, band pass, and band reject filters by straight-forward transformations. These transformations will be covered in Section 8.2, after a discussion of selected low pass filter circuit examples.

8.2 PASSIVE FILTERS

We now turn to examples of Butterworth, Chebyshev and elliptic low pass filter designs, normalized to a 50 ohm input and output impedance and a cutoff frequency of 1 MHz.

Figure 8-5 illustrates a capacitive input Butterworth low pass filter. In Figure 8-5, the following element values result in a filter with a 3 dB cutoff frequency of 1 MHz and an input and output impedance of 50 ohms:

$$C_1 = 1965 \text{ pF} \qquad L_2 = 12.9 \ \mu\text{H}$$

$$C_3 = 6360 \text{ pF} \qquad L_4 = 12.9 \ \mu\text{H}$$

$$C_5 = 1965 \text{ pF}$$

In some circuit designs, it is desirable to utilize a filter with an inductive input to minimize the possibility of oscillation in an active device. Figure 8-6 illustrates an inductive input 5-element Butterworth low pass filter. In Figure 8-6, the following element values result in a filter with a 3 dB cutoff frequency of 1 MHz and an input and output impedance of 50 ohms:

$$C_2 = 5150 \text{ pF} \qquad L_1 = \ 4.9 \ \mu\text{H}$$

$$C_4 = 5150 \text{ pF} \qquad L_3 = 15.9 \ \mu\text{H}$$

$$L_5 = \ 4.9 \ \mu\text{H}$$

Figure 8-7 illustrates a capacitive input, 5-element Chebyshev low pass filter. Note that the network configuration of the Chebyshev filter is identical to that of the Butterworth filter. The two filter types differ only in their element values. The

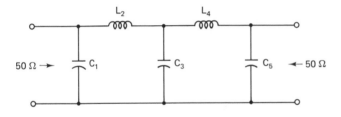

Figure 8-5 Five element capacitive input Butterworth low pass filter.

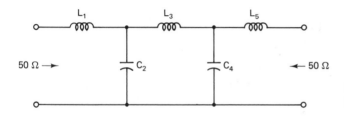

Figure 8-6   Five element inductive input Butterworth low pass filter.

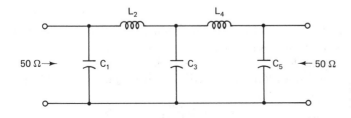

Figure 8-7   Five element capacitive input Chebyshev low pass filter.

following element values result in a filter with a 3 dB cutoff frequency of 1 MHz and an input and output impedance of 50 ohms:

$$C_1 = 3627 \text{ pF} \quad L_2 = 13.0 \text{ } \mu\text{H}$$

$$C_3 = 6770 \text{ pF} \quad L_4 = 13.0 \text{ } \mu\text{H}$$

$$C_5 = 3627 \text{ pF}$$

A 7-element Chebyshev low pass filter with capacitive input is illustrated in Figure 8-8.   In Figure 8-8, the following element values result in a filter with a 3 dB cutoff frequency of 50 ohms and an input and output impedance of 50 ohms:

$$C_1 = 3137 \text{ pF} \quad L_2 = 12.7 \text{ } \mu\text{H}$$

$$C_3 = 6507 \text{ pF} \quad L_4 = 14.6 \text{ } \mu\text{H}$$

$$C_5 = 6507 \text{ pF} \quad L_6 = 12.7 \text{ } \mu\text{H}$$

$$C_7 = 3137 \text{ pF}$$

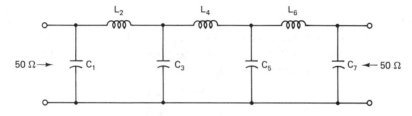

Figure 8-8   Seven element capacitive input Chebyshev low pass filter.

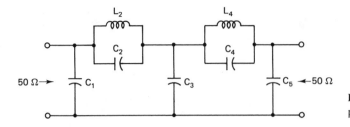

Figure 8-9 Fifth degree elliptic low pass filter.

The 7-element filter has a sharper cutoff characteristic than the 5-element filter, but has a larger number of ripples in its passband.

Figure 8-9 illustrates a fifth degree elliptic low pass filter. In Figure 8-9, the following element values result in a filter with a 3 dB cutoff frequency of 1 MHz and an input and output impedance of 50 ohms:

$$C_1 = 2943 \text{ pF} \qquad L_2 = 11.6 \text{ μH}$$

$$C_3 = 5559 \text{ pF} \qquad L_4 = 9.5 \text{ μH}$$

$$C_5 = 2398 \text{ pF}$$

$$C_2 = 363 \text{ pF}$$

$$C_4 = 1046 \text{ pF}$$

An example filter may easily be scaled to a different cutoff frequency or impedance level. If a filter is desired with a cutoff frequency of n times the 1 MHz cutoff of a given filter, each value of inductance and capacitance is divided by the factor n. If an impedance level of m times the 50 ohm level of the given filters is desired, each inductor value must be multiplied by m and each capacitor value must be divided by m.

Example 8-1

Design a 5-element capacitive input Butterworth low pass filter with a cutoff frequency of 5 MHz and an impedance of 72 ohms.

Solution

The 50 ohm, 1 MHz filter is given in Figure 8-10. The frequency transformation factor for the new filter is 5 and the impedance transformation factor is 1.44. Each inductance value must be multiplied by

$$L' = \frac{mL}{n} = \frac{(1.44)L}{(5)} = 0.29L$$

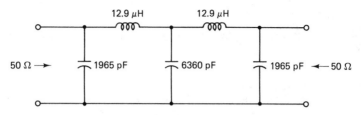

Figure 8-10   Filter for Example 8-1.

Each capacitance value must be multiplied by:

$$C' = \frac{C}{mn} = \frac{C}{(1.44)(5)} = 0.14C$$

The transformed filter with a cutoff frequency of 5 MHz and a 72 ohm impedance level is illustrated in Figure 8-11.

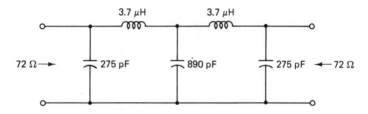

Figure 8-11   Transformed filter for Example 8-1.

A low pass filter may be tranformed into a band pass filter with center frequency f_o and bandwidth BW with a slightly more complex transformation. To transform, each inductor in the low pass filter must be replaced by a capacitor and inductor in series, and each capacitor must be replaced by a capacitor and inductor in parallel. Figure 8-12 illustrates the transformation and gives the expressions necessary to compute the band pass filter component values from the low pass filter component values.

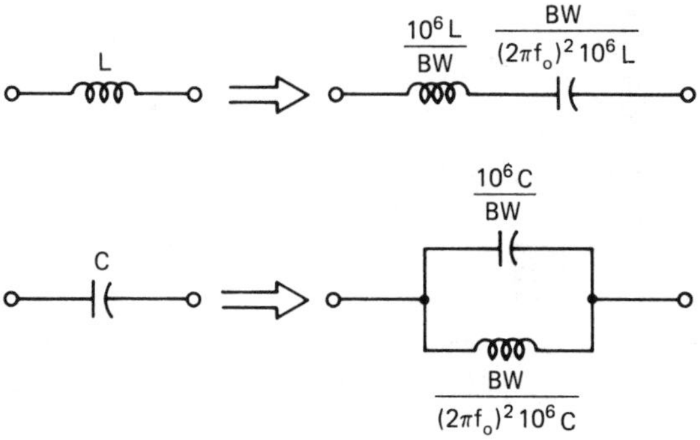

Figure 8-12   Low pass filter to band pass filter transformations.

Example 8-2

Design a 10-element, 50 ohm Butterworth band pass filter whose center frequency is 10 MHz and whose bandwidth is 2 MHz.

Solution

The 5-element capacitive input Butterworth low pass filter will be transformed into the desired band pass filter. The filter configuration is illustrated in Figure 8-13.

$$L_1 = L_5 = \frac{BW}{(2\pi)^2 f_o^2 10^6 C}$$

$$= \frac{2 \times 10^6}{(6.28)^2 (10 \times 10^6)^2 (10^6)(1965 \times 10^{-12})} = 0.26 \ \mu H$$

$$C_1 = C_5 = \frac{10^6 C}{BW} = \frac{(10^6)(1965 \times 10^{-12})}{(2 \times 10^6)} = 983 \ pF$$

$$L_3 = \frac{BW}{(2\pi)^2 f_o^2 10^6 C}$$

$$= \frac{2 \times 10^6}{(6.28)^2 (10 \times 10^6)^2 (10^6)(6360 \times 10^{-12})} = 0.08 \ \mu H$$

$$C_3 = \frac{10^6 C}{BW} = \frac{(10^6)(6360 \times 10^{-12})}{2 \times 10^6} = 3180 \ pF$$

$$C_2 = C_4 = \frac{BW}{(2\pi)^2 f_o^2 \ 10^6 L}$$

$$= \frac{2 \times 10^6}{(6.28)^2 (10 \times 10^6)^2 (10^6)(12.9 \times 10^{-6})} = 39.3 \ pF$$

$$L_2 = L_4 = \frac{10^6 L}{BW} = \frac{(10^6)(12.9 \times 10^{-6})}{2 \times 10^6} = 6.5 \ \mu H$$

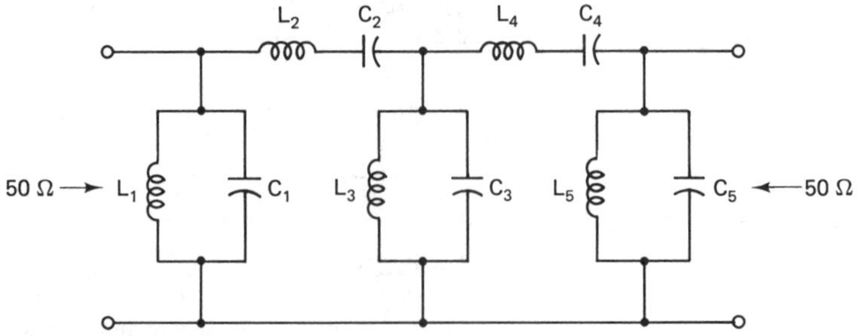

Figure 8-13   Filter configuration for Example 8-2.

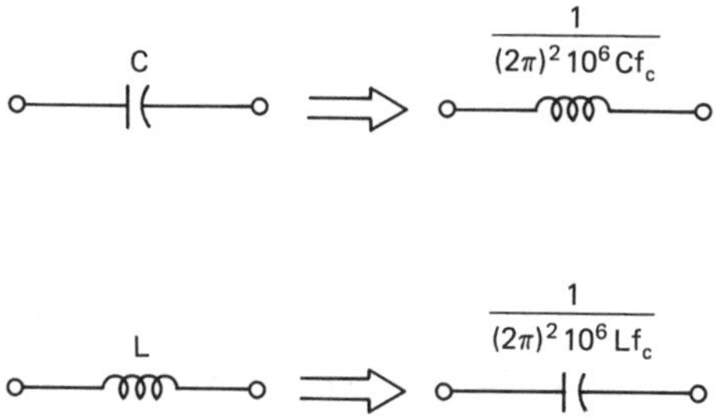

$$\frac{1}{(2\pi)^2 \, 10^6 \, Cf_c}$$

C

$$\frac{1}{(2\pi)^2 \, 10^6 \, Lf_c}$$

L

Figure 8-14 Low pass filter to high pass filter transformations.

The transformation from a low pass filter to a high pass filter is very straightforward. Inductors in the low pass filter become capacitors in the high pass filter, and capacitors in the low pass filter become inductors in the high pass configuration. Figure 8-14 illustrates the necessary transformations.

Example 8-3

Design a 50 ohm, five-element Chebyshev high pass filter with a 3 dB cutoff frequency of 8 MHz.

Solution

The low pass to high pass transformation yields the filter configuration illustrated in Figure 8-15.

$$C_2 = C_4 = \frac{1}{(2\pi)^2 10^6 L f_c}$$

$$= \frac{1}{(6.28)^2 (10^6)(13 \times 10^{-6})(8 \times 10^6)} = 244 \text{ pF}$$

$$L_1 = L_5 = \frac{1}{(2\pi)^2 10^6 C f_c}$$

$$= \frac{1}{(6.28)^2 (10^6)(3627 \times 10^{-12})(8 \times 10^6)} = 0.87 \text{ } \mu\text{H}$$

$$L_3 = \frac{1}{(2\pi)^2 10^6 C f_c}$$

$$= \frac{1}{(6.28)^2 (10^6)(6770 \times 10^{-12})(8 \times 10^6)} = 0.47 \text{ } \mu\text{H}$$

A low pass filter may be transformed into a band reject filter with center frequency f_o and bandwidth BW with a transformation similar in complexity to the

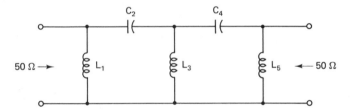

Figure 8-15 Filter configuration for Example 8-3.

band pass transformation. To transform, each inductor in the low pass filter must be replaced by a capacitor and inductor in parallel, and each capacitor in the low pass filter must be replaced by an inductor in series with a capacitor. Figure 8-16 illustrates the transformation.

Example 8-4

Design a 10-element, 50 ohm Butterworth band reject filter whose center frequency is 10 MHz and whose bandwidth is 2 MHz.

Solution

The 5-element capacitive input Butterworth low pass filter will be transformed into the desired band reject filter. The filter configuration is illustrated in Figure 8-17.

$$L_1 = L_5 = \frac{1}{(2\pi)^2 10^6 BW\ C}$$

$$= \frac{1}{(6.28)^2 (10^6)(2 \times 10^6)(1965 \times 10^{-12})} = 6.5\ \mu H$$

$$C_1 = C_5 = \frac{10^6 BW\ C}{f_o^2}$$

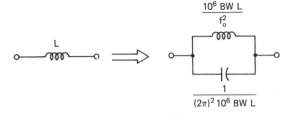

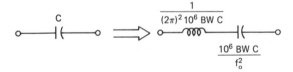

Figure 8-16 Low pass filter to band reject filter transformation.

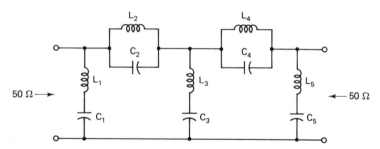

Figure 8-17   Filter configuration for Example 8-4.

$$= \frac{(10^6)(2 \times 10^6)(1965 \times 10^{-12})}{(10 \times 10^6)^2} = 39.3 \text{ pF}$$

$$L_3 = \frac{1}{(2\pi)^2 10^6 BW\ C}$$

$$= \frac{1}{(6.28)^2(10^6)(2 \times 10^6)(6360 \times 10^{-12})} = 2.0\ \mu\text{H}$$

$$C_3 = \frac{10^6 BW\ C}{f_o^2}$$

$$= \frac{(10^6)(2 \times 10^6)(6360 \times 10^{-12})}{(10 \times 10^6)^2} = 127 \text{ pF}$$

$$L_2 = L_4 = \frac{10^6 BW\ L}{f_o^2}$$

$$= \frac{(10^6)(2 \times 10^6)(12.9 \times 10^{-6})}{(10 \times 10^6)^2} = 0.26\ \mu\text{H}$$

$$C_2 = C_4 = \frac{1}{(2\pi)^2 10^6 BW\ L}$$

$$= \frac{1}{(6.28)^2(10^6)(2 \times 10^6)(12.9 \times 10^{-6})} = 983 \text{ pF}$$

It is important to note that the transformation equations given in this section apply only to low pass filters that are normalized to 1 MHz and 50 ohms. Some filter tables are normalized to 1 radian per second and/or 1 ohm impedance. Use the transformation relationships given in that table or first transform the filter to 1 MHz and 50 ohms before using the equations given here.

It is also important to realize that this section has presented only examples of possible filters. Many filter designs are possible depending upon the characteristics (such as pass band ripple magnitude, etc.) desired. Filter tables should be consulted for designs where these characteristics must be specified. All filters discussed in this section have a ripple amplitude of 0.1 dB or less.

Filter tables are available that provide component values in terms of standard values of capacitance. If such tables are not available, the calculated capacitance values must be realized by choosing the nearest standard value, or capacitors may be wired in parallel to achieve the desired value. Inductors often may be wound to the exact value required.

Lumped filters, such as those described in this section, may be used up to frequencies of a few GHz; above this range, distributed filters constructed from transmission lines are generally used. Computer-aided filter design software, such as E-Syn, developed by EEsof of Westlake Village, California, is available to design both lumped and distributed filters, given desired filter specifications. E-syn is available for the IBM PC and compatibles as well as for more powerful computers, such as the DEC VAX series.

8.3 ACTIVE FILTERS

Active filters are implemented with operational amplifiers, resistors and capacitors (inductors are normally avoided since the availability of signal inversion in the operational amplifier allows a capacitor to appear as an inductor). A number of active filter examples will be presented in this section; many other implementations are possible—active filter handbooks should be consulted for more complete information. Active filters are limited in frequency by the characteristics of the operational amplifier used in the circuit and thus are most useful for low frequency filtering applications. Active filters are also limited in their power handling capability and thus are useful only for low level signal processing. A unique characteristic of active filters is the possibility of a gain greater than 1.0 which, of course, is not possible with a passive filter.

One of the simplest inverting gain active filter circuits is the infinite gain, multiple feedback (MFB) circuit illustrated in Figure 8-18. This circuit may be used to implement a second order Chebyshev filter depending upon the circuit constants chosen. This circuit may be cascaded with similar second order filters to achieve a filter of order four, six, or higher (even order) or may be cascaded with a first order filter to achieve an odd order filter.

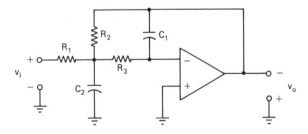

Figure 8-18 Infinite gain multiple feedback low pass active filter.

Second order MFB low pass filters with 3 dB cutoff frequency, f_c, may be designed as follows:

1. Determine the desired gain K and the filter type (Butterworth or Chebyshev), which determines the constants B and C.
2. Select a standard value for C_2 of approximately $10/f_c$ µF.
3. Select a standard value for C_1 to satisfy the relation

$$C_1 \leq \frac{B^2 C_2}{4C(K + 1)}$$

4. Calculate R_2 from

$$R_2 = \frac{2(K + 1)}{[BC_2 + \sqrt{B^2 C_2^2 - 4CC_1 C_2(K + 1)}]2\pi f_c}$$

5. Calculate R_1 from

$$R_1 = \frac{R_2}{K}$$

6. Calculate R_3 from

$$R_3 = \frac{1}{CC_1 C_2 (2\pi f_c)^2 R_2}$$

For a second order Butterworth filter, the constants B and C are $B = 1.414$ and $C = 1.000$. For a second order Chebyshev filter with passband ripple width of 0.1 dB, $B = 2.372$ and $C = 3.314$.

Example 8-5

Design an inverting second order Butterworth low pass active filter with a gain of 5 and a cutoff frequency of 2 KHz.

Solution

1. $B = 1.414$ $C = 1.000$ $K = 5$
2. $C_2 \approx 10/f_c$ µF $= 10/2000$ µF $= 0.005$ µF.   Choose $C_2 = 5100$ pF.
3. $C_1 \leq \dfrac{B^2 C_2}{4C(K + 1)} = \dfrac{(1.414)^2 (0.0051 \times 10^{-6})}{(4)(1)(5 + 1)} = 425$ pF.
   Choose $C_1 = 390$ pF.

   $$R_2 = \frac{2(K + 1)}{[BC_2 + \sqrt{B^2 C_2^2 - 4CC_1 C_2(K + 1)}]\, 2\pi f_c}$$

$$4. \quad = \frac{2(5 + 1)}{[(1.414)(5100 \times 10^{-12}) + \sqrt{(1.414)^2(5100 \times 10^{-12})^2 - (4)(1)(390 \times 10^{-12})(5100 \times 10^{-12})(5 + 1)}] \, (6.28)(2 \times 10^3)}$$

$$= 103 \text{ K}$$

$$5. \; R_1 = \frac{R_2}{K} = \frac{103 \text{ K}}{6} = 17.1 \text{ K}$$

$$6. \; R_3 = \frac{1}{CC_1C_2(2\pi f_c)^2 R_2}$$

$$R_3 = \frac{1}{(1)(390 \times 10^{-12})(5100 \times 10^{-12})(6.28)^2(2 \times 10^3)^2(103 \times 10^3)}$$

$$= 30.9 \text{ K}$$

Choose standard resistance values as follows: $R_1 = 18$ K; $R_2 = 100$ K; $R_3 = 30$ K.

The filter circuit is given in Figure 8-19.

A fourth order Butterworth filter may be implemented by cascading two second order stages with the following values for the constants B, C and K:

Stage 1	Stage 2	Overall filter gain
Stage Gain = K_1	Stage Gain = K_2	$K = K_1 K_2$
$B_1 = 0.765$	$B_1 = 1.848$	
$C_1 = 1.000$	$C_1 = 1.000$	

A fourth order Chebyshev filter with a pass band ripple width of 0.1 dB may also be implemented by cascading two second order stages. The following constants apply:

Stage 1	Stage 2	Overall filter gain
Stage Gain = K_1	Stage Gain = K_2	$K = K_1 K_2$
$B_1 = 0.528$	$B_2 = 1.275$	
$C_1 = 1.330$	$C_2 = 0.623$	

Constants for odd order and higher even order filters are available in active filter handbooks.

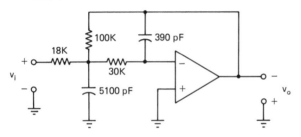

Figure 8-19   Filter circuit for Example 8-5.

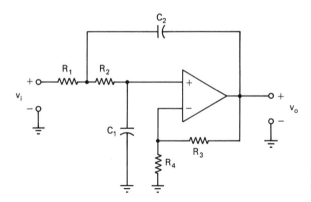

Figure 8-20 Voltage controlled voltage source active low pass filter.

If an active filter with a non-inverting gain ($K \geq 1$) is required, the voltage controlled voltage source (VCVS) configuration illustrated in Figure 8-20 is widely used. Second order VCVS low pass filters with a 3 dB cutoff frequency f_c may be designed as follows:

1. Determine the desired gain K and the filter type (Butterworth or Chebyshev), which determines the constants B and C.
2. Select a standard value for C_2 of approximately $10/f_c$ μF.
3. Select a standard value for C_1 that satisfies the relation

$$C_1 \leq \frac{[B^2 + 4C(K - 1)]C_2}{4C}$$

4. Calculate R_1 from

$$R_1 = \frac{2}{[BC_2 + \sqrt{\{B^2 + 4C(K - 1)\}C_2^2 - 4CC_1C_2}] \, 2\pi f_c}$$

5. Calculate R_2 from

$$R_2 = \frac{1}{CC_1C_2R_1(2\pi f_c)^2}$$

6. Calculate R_3 from

$$R_3 = \frac{K(R_1 + R_2)}{K - 1}$$

   If $K = 1$, $R_3 = \infty$ (open)
7. Calculate $R_4 = K(R_1 + R_2)$.
   If $K = 1$, $R_4 = 0$ (short)

The constants B and C in these expressions are the same as given for the MFB filter configuration.

Example 8-6

Design a non-inverting second order Chebyshev low pass active filter with a gain of 10, a cutoff frequency of 1 KHz, and a pass band ripple width of 0.1 dB.

Solution

1. $B = 2.372$ $C = 3.314$ $K = 10$
2. $C_2 \approx 10/f_c$ $\mu F = 10/1000$ $\mu F = 0.01$ μF
   Choose $C_2 = 0.01$ μF.
3. $C_1 \leq \dfrac{[B^2 + 4C(K - 1)]C_2}{4C}$

   $C_1 \leq \dfrac{[(2.372)^2 + (4)(3.314)(10 - 1)] \, [0.01 \times 10^{-6}]}{(4)(3.314)}$

   $C_1 \leq 0.094$ μF
   Choose $C_1 = 0.091$ μF.

4. $R_1 = \dfrac{2}{[BC_2 + \sqrt{\{B^2 + 4C(K - 1)\}C_2^2 - 4CC_1C_2}] \, 2\pi f_c}$

   $= \dfrac{2}{\begin{array}{c}[(2.372)(0.01 \times 10^{-6}) + \sqrt{\{(2.372)^2 + (4)(3.314)(9)\}} \\ \overline{(0.01 \times 10^{-6})^2 - (4)(3.314)(0.09 \times 10^{-6})(0.01 \times 10^{-6})}](6.28 \times 10^3)\end{array}}$

   $= 7.3$ K

5. $R_2 = \dfrac{1}{CC_1C_2R_1 \, (2\pi f_c)^2}$

   $= \dfrac{1}{(3.314)(0.091 \times 10^{-6})(0.01 \times 10^{-6})(7.3 \times 10^3)(6.28)^2 \, 10^6}$
   $= 1.2$ K

6. $R_3 = \dfrac{K(R_1 + R_2)}{K - 1} = \dfrac{10(1.2 \times 10^3 + 7.3 \times 10^3)}{9} = 9.4$ K

7. $R_4 = K(R_1 + R_2) = 10(7.3 \times 10^3 + 1.2 \times 10^3) = 85$ K
   Choose standard resistance values as follows:

$$R_1 = 7.5 \text{ K} \qquad R_3 = 9.1 \text{ K}$$

$$R_2 = 1.2 \text{ K} \qquad R_4 = 8.2 \text{ K}$$

The filter circuit is illustrated in Figure 8-21.

Inverse Chebyshev and elliptic low pass filter designs are available in active filter handbooks. An excellent filter configuration that exhibits greater stability and higher Q (quality factor − a measure of selectivity) than the MFB or VCVS configuration, but which requires three operational amplifiers for a second order

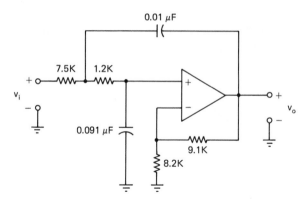

Figure 8-21   Filter circuit for Example 8-6.

filter, is the biquad configuration.   Filter design information for the biquad configuration is also available in filter handbooks and will not be covered here.

A Butterworth or Chebyshev high pass filter may be realized with the infinite gain multiple feedback configuration, as illustrated in Figure 8-22.

Second order MFB high pass filters may be designed as follows:

1. Determine the desired gain K and filter type (Butterworth or Chebyshev), which determines the constants B and C (same constants as for low pass filters).
2. Select a standard value for C_1 of approximately $10/f_c$ μF.
3. Calculate C_2 from:

$$C_2 = \frac{C_1}{K}$$

4. Calculate R_1 from:

$$R_1 = \frac{B}{(2C_1 + C_2)\,2\pi f_c}$$

5. Calculate R_2 from:

$$R_2 = \frac{(2C_1 + C_2)C}{BC_1 C_2\,2\pi f_c}$$

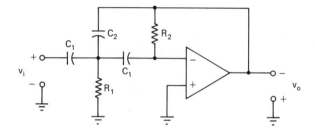

Figure 8-22   Infinite gain multiple feedback high pass active filter.

Example 8-7

Design an inverting second order Chebyshev high pass active filter with a gain of 10, a cutoff frequency of 1 KHz, and a pass band ripple width of 0.1 dB.

Solution

1. $B = 2.372$ $C = 3.314$ $K = 10$

2. $C_1 \approx 10/f_c$ $\mu F = 10/1000$ $\mu F = 0.01$ μF
   Choose $C_1 = 0.01$ μF.

3. $C_2 = \dfrac{C_1}{K}$

   $C_2 = \dfrac{0.01 \times 10^{-6}}{10} = 0.001 \times 10^{-6}$

   $C_2 = 0.001$ μF

4. $R_1 = \dfrac{B}{(2C_1 + C_2)\, 2\pi f_c}$

   $= \dfrac{2.372}{[(2)(0.01 \times 10^{-6}) + (0.001 \times 10^{-6})]\,[6.28]\,[10^3]}$

   $= 18\text{K}$

5. $R_2 = \dfrac{(2C_1 + C_2)C}{BC_1C_2\, 2\pi f_c}$

   $= \dfrac{[(2)(0.01 \times 10^{-6}) + (0.001 \times 10^{-6})][3.314]}{(2.372)(0.01 \times 10^{-6})(0.001 \times 10^{-6})(6.28)(10^3)}$

   $= 467\text{ K}$

The circuit is illustrated in Figure 8-23.

A second order band pass filter may be realized with the MFB configuration as illustrated in Figure 8-24.

Second order MFB band pass filters may be designed as follows:

1. Select a standard value of C_1 of approximately $10/f_o$ μF.

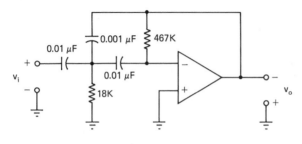

Figure 8-23   Filter circuit for Example 8-7.

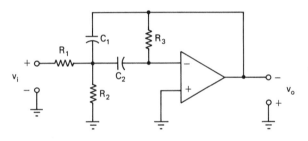

Figure 8-24 Infinite gain multiple feedback band pass active filter.

2. Select a standard value of C_2 from

$$C_2 > C_1 \left[\frac{K}{Q^2} - 1 \right]$$

3. Calculate R_1 from

$$R_1 = \frac{1}{\left[\dfrac{K}{Q} \right] 2\pi f_o C_1}$$

4. Calculate R_2 from

$$R_2 = \frac{1}{\left[C_1 \left(1 - \dfrac{K}{Q^2} \right) + C_2 \right] 2\pi f_o Q}$$

5. Calculate R_3 from

$$R_3 = \frac{Q}{2\pi f_o} \left[\frac{1}{C_1} + \frac{1}{C_2} \right]$$

Note that quality factor is given by $Q = \dfrac{f_o}{BW}$.

Example 8-8

Design an inverting second order bandpass active filter with a center frequency of 10 KHz, a bandwidth of 1 KHz, and a gain of 5.

Solution

1. $C_1 \approx 10/f_o$ $\mu F = 10/10000$ $\mu F = 0.001\ \mu F$
   Choose $C_1 = 0.001\ \mu F$.

2. $Q = \dfrac{f_o}{BW} = \dfrac{10 \times 10^3}{1 \times 10^3} = 10$

$$C_2 > C_1 \left[\dfrac{K}{Q^2} - 1 \right]$$

$$C_2 > C_1 \left[\dfrac{5}{100} - 1 \right] > -0.95 C_1$$

$\therefore$ any positive value of C_2 is acceptable. Choose $C_2 = 0.01 \ \mu\text{F}$.

3. $R_1 = \dfrac{1}{\left(\dfrac{K}{Q} \right) 2\pi f_o C_1}$

$= \dfrac{1}{\left(\dfrac{5}{10} \right)(2\pi)(10 \times 10^3)(0.001 \times 10^{-6})}$

$= 31.8 \ \text{K}$

4. $R_2 = \dfrac{1}{\left[C_1 \left(1 - \dfrac{K}{Q^2} \right) + C_2 \right] 2\pi f_o Q}$

$= \dfrac{1}{[(0.001 \times 10^{-6})\left(1 - \dfrac{5}{100}\right) + 0.01 \times 10^{-6}][6.28][10 \times 10^3][10]}$

$= 145 \ \Omega$

5. $R_3 = \dfrac{Q}{2\pi f_o} \left[\dfrac{1}{C_1} + \dfrac{1}{C_2} \right]$

$= \dfrac{10}{(6.28)(10 \times 10^3)} \left[\dfrac{1}{0.001 \times 10^{-6}} + \dfrac{1}{0.01 \times 10^{-6}} \right]$

$= 175 \ \text{K}$

Band reject filters may be implemented using circuits similar to those presented in this section. Readers are again referred to active filter handbooks for details.

8.4 SWITCHED CAPACITOR FILTERS

It is extremely desirable to implement often-used filter functions in integrated circuit form. Integrated circuit manufacturers have attempted to integrate active RC filters such as those discussed in Section 8.3 on a single chip, but have achieved only limited success. Active filters require accurate values of resistance and ca-

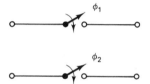

Figure 8-25 Switch representation.

pacitance for satisfactory performance; this is very difficult to achieve in integrated circuits without expensive trimming. An additional problem with integrated resistors is that they have very poor linearity and temperature characteristics. In addition to all of these problems, large time constants require large values of capacitance and resistance, which in turn require large areas on the integrated circuit. A solution to the many problems indicated above is to replace the resistors in the integrated circuit by switches and capacitors. In such a circuit, performance is a function of capacitor ratios rather than capacitor absolute values. Capacitor ratios are much easier to control in integrated circuits. Filters implemented with switched capacitors replacing resistors are known as switched capacitor filters.

Figure 8-25 illustrates the notation used for switches in switched capacitor circuits. In Figure 8-25, as in all switched capacitor circuit diagrams, the switches are shown open. The symbols ϕ_1 and ϕ_2 indicate that the switches close during phases ϕ_1 and ϕ_2 respectively. ϕ_1 and ϕ_2 are determined by a non-overlapping phase clock as illustrated in Figure 8-26. In the notation to be used in this section, the switches close when the clock pulse ϕ_1 or ϕ_2 is high. Note that the non-overlapping property causes both switches to remain open for a finite period of time between pulses. A more complex three-switch circuit would require a three-phase clock, a four-switch circuit a four-phase clock, and so on.

Figure 8-27 illustrates the *parallel switched capacitor resistor realization*. In the switched capacitor circuit in Figure 8-27, the switches controlled by ϕ_1 and ϕ_2

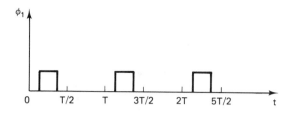

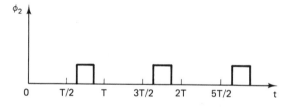

Figure 8-26 Non-overlapping two phase clock.

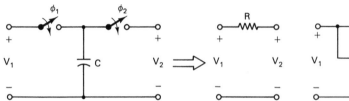

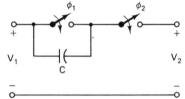

Figure 8-27 Parallel switched capacitor resistor realization.

Figure 8-28 Series switched capacitor resistor realization.

alternately close and open, resulting in an average current through the upper branches of the circuit given by

$$I_{av} = (V_1 - V_2)\left(\frac{C}{T}\right)$$

By Ohm's law, the equivalent resistance of the switched capacitor circuit is given by

$$R = \frac{T}{C}$$

Figure 8-28 illustrates the series switched capacitor resistor realization. The equivalent resistance of the series switched capacitor resistor realization is also $R = T/C$. Although the equivalent resistance is the same for the parallel and series realizations, the two circuits differ in that the output is isolated from the input between clock pulses in the parallel configurations, while the output and input are connected through the capacitor between clock pulses in the series realization. This difference is important for some circuits.

Figure 8-29 illustrates a third realization, the series-parallel configuration. The equivalent resistance for this circuit is

$$R = \frac{T}{2C}$$

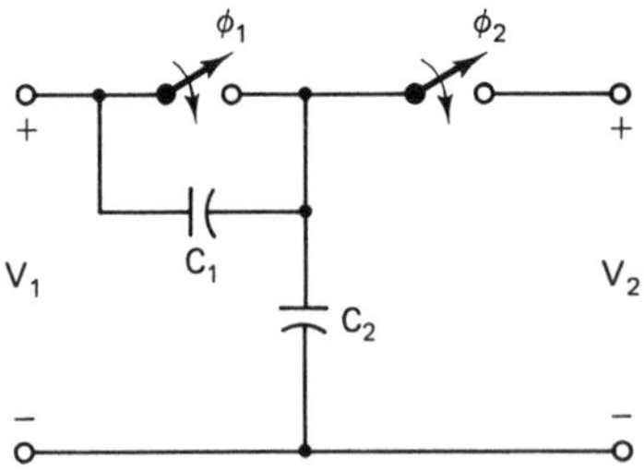

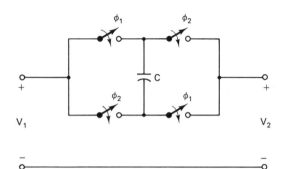

Figure 8-29 Series-parallel switched capacitor resistor realization.

Figure 8-30 Bilinear switched capacitor resistor realization.

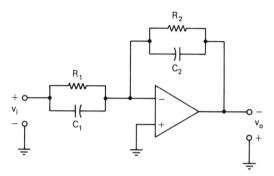

Figure 8-31   First order active filter for Example 8-9.

The bilinear switched capacitor realization is illustrated in Figure 8-30. The equivalent resistance for this circuit is

$$R = \frac{T}{4C}$$

Example 8-9

Implement the first-order active filter illustrated in Figure 8-31 as a switched capacitor filter.

Solution

The parallel switched capacitor resistor realization will be used and is illustrated in Figure 8-32. Note that ϕ_1 and ϕ_2 are reversed in the feedback network because of the inverting operational amplifier connection. The capacitor values C_3 and C_4 are given by:

$$C_3 = \frac{T}{R_1} \qquad C_4 = \frac{T}{R_2}$$

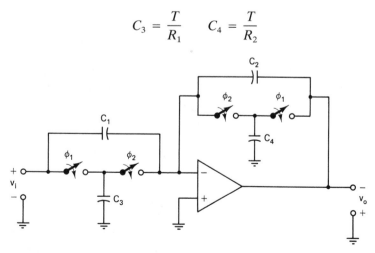

Figure 8-32   Parallel switched capacitor resistor realization for Example 8-9.

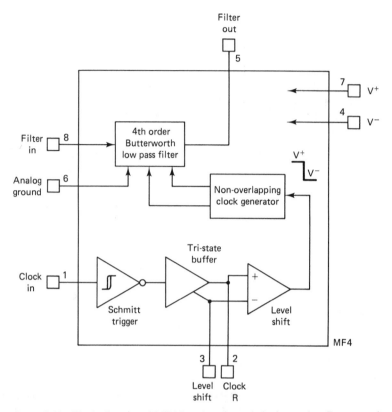

Figure 8-33   Block diagram of MF4 fourth order switched capacitor Butterworth low pass filter (reprinted with permission of National Semiconductor Corporation).

The switches shown in the switched capacitor circuits illustrated in this section are normally metal oxide semiconductor (MOS) solid state switches, which are integrated onto the same substrate as the capacitors and operational amplifier circuitry.

Figure 8-33 illustrates the block diagram of the National Semiconductor MF4 fourth order switched capacitor Butterworth low pass filter, available in integrated circuit form. The MF4 is available in two versions, the MF4-50 and the MF4-100. The cutoff frequency of the filter is determined by the external clock frequency. The ratio of clock frequency to cutoff frequency is 50:1 in the MF4-50 and 100:1 in the MF4-100. The device operates from a supply voltage of ±5 volts to ±14 volts and has a cutoff frequency range of 0.1 Hz to 20 KHz. The MF4 can be driven by TTL or CMOS level clock signals or can operate with an internal clock with the addition of an external resistor and capacitor. MF4s can be cascaded to realize higher order Butterworth filters.

Other switched capacitor filters that realize the high pass, low pass, and bandpass functions in both Butterworth and Chebyshev type filters are available in integrated circuit form. The National Semiconductor MF10 is a universal switched capacitor filter that can realize any of the above filter functions and/or types by selecting the correct output.

8.5 DIGITAL FILTERS

A digital filter is a signal processing unit that converts an input sequence of binary digits into an output sequence of binary digits. The output sequence is a filtered version of the input sequence if the signal processing unit performs an appropriate combinations of delays, sums, and multiplications on this input sequence. Figure 8-34 illustrates the concept of digital filtering.

In Figure 8-34, a continuous analog input signal $v_i(t)$ is converted to a discrete-time digital sequence, $v_i(nT)$, by an analog-to-digital converter. The output of the analog-to-digital converter is applied to the input of the digital filter, which modifies the sequence to $v_o(nT)$. A digital-to-analog converter converts the filter output sequence to the continuous analog output signal $v_o(t)$.

Digital filters may be implemented in software, most often with a 16- or 32-bit microprocessor and an associated coprocessor chip or entirely in hardware with adders, shift registers (delay elements), and multipliers utilized individually or as functions in a dedicated signal processing chip.

Figure 8-35 illustrates the three basic building blocks for a digital filter: the delay unit, the summer, and the multiplier. The delay unit is in effect a storage unit that can be implemented in hardware with a shift register or in software in a storage register or memory; the symbol z^{-1} in the delay unit block arises from a technique used in the analysis and design of digital filters known as the z-transform. The summer unit simply provides an output equal to the sum of its inputs and can be realized with a hardware adder or in software. The multiplier unit scales the input by a factor α; it can also be easily implemented in hardware or software.

Digital filters can be designed to implement the low pass, high pass, band pass and band reject functions with the standard response characteristics, such as Butterworth or Chebyshev. Digital filter design is beyond the scope of this text. A detailed knowledge of design techniques, however, is not required for use of digital filters as integrated signal processing units in communications systems. Figure 8-36 illustrates a simple analog RC low pass filter and its digital counterpart to give the reader a feel for digital filter realization. In Figure 8-36, T is the sampling period of the analog-to-digital and digital-to-analog converters, and R and C are the resistance and capacitance values in the analog circuit.

An example of a commercially available digital filter in integrated circuit form

Figure 8-34   Digital filtering concept.

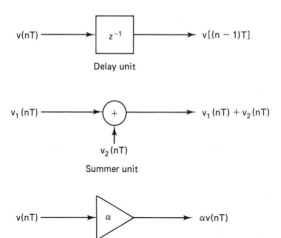

Figure 8-35 Digital filter building blocks.

is the Plessey MS2003 digital filter and detector circuit. This chip contains eight second order recursive digital filter sections, which may be used as building blocks to realize higher order digital filters. The chip also contains level detection circuitry. The digital filter constants in the MS2003 are determined by data stored in an external memory chip, such as a programmable read only memory (PROM) or a random access memory (RAM). An input scaler is included to prevent overflow in the filter, since many digital filters exhibit a gain of greater than unity.

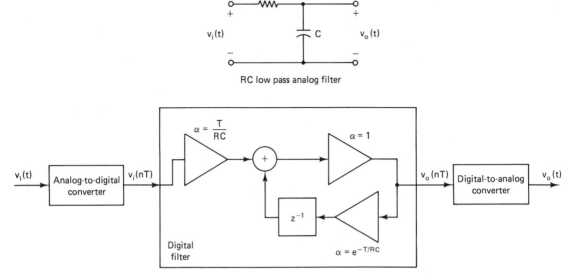

Figure 8-36 Digital filter realization of RC low pass analog filter.

8.6 RF TRANSFORMERS

A transformer converts an impedance from one level to another using mutual magnetic coupling. Transformers are used in communications systems at power and audio frequencies, as well as in the RF portion of the frequency spectrum. This section will discuss transformer circuits useful for impedance transformation applications at RF frequencies.

Figure 8-37 illustrates the schematic diagram of an air-core transformer. In Figure 8-37, L_p and L_s are the primary and secondary inductances respectively, V_p and V_s are the primary and secondary voltages, I_p and I_s are the primary and secondary currents, M is the mutual inductance between the primary and secondary windings, and the dots indicate the polarity of the mutual coupling. An increasing current flowing into the dot on one winding will introduce a positive voltage at the dot on the other winding. A measure of the magnetic flux linking the two coils is the coefficient of coupling k. The coefficient of coupling is 1.0 when all of flux from one coil links the others and 0.0 when there are no flux linkages. The mutual inductance of a transformer is related to the coefficient of coupling by

$$M = k \sqrt{L_p L_s}$$

The coefficient of coupling and thus the mutual inductance may be increased by winding the transformer on a high permeability core as opposed to an air core. The high permeability core, if properly selected for the frequency range of interest, can allow the coefficient of coupling to approach 1.0. Ferrite materials are often used for cores at RF frequencies. If the coefficient of coupling is approximately unity and losses such as winding resistance losses, hysteresis and eddy current losses in the core, and the leakage reactance (flux lines that do not link both windings) are negligible, the transformer may be considered ideal and its analysis simplified greatly.

If an ideal transformer has N_p turns on its primary coil and N_s turns on its secondary coil, the relationship between the primary and secondary voltages is

$$\frac{V_p}{V_s} = \frac{N_p}{N_s}$$

The relationship between primary and secondary currents is

$$\frac{I_p}{I_s} = \frac{N_s}{N_p}$$

Note that if it is assumed that primary and secondary voltages and currents are RMS values, the primary power $V_p I_p$ equals the secondary power $V_s I_s$. This is a result of assuming an ideal transformer with no losses.

Figure 8-38 illustrates an ideal transformer whose secondary is terminated

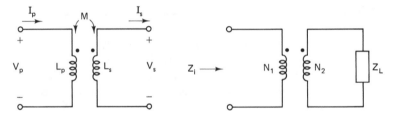

Figure 8-37 Schematic diagram of an
air-core transformer.

Figure 8-38 Transformer terminated
with complex impedance.

with a complex impedance Z_L. In Figure 8-38, the input impedance, Z_i, is given
by:

$$Z_i = \left(\frac{N_p}{N_s}\right)^2 Z_L$$

The ideal transformer may therefore be used as an impedance transformation circuit
by appropriately adjusting the turns ratio N_p/N_s.

Example 8-10

A 10 MHz RF signal is applied to an ideal transformer, whose secondary is terminated by a load of 50 ohms in series with 1000 pf. The turns ratio of the
transformer is 3:1. Calculate the impedance looking into the input of the transformer.

Solution

$$X_{cL} = \frac{1}{2\pi f C} = \frac{1}{(6.28)(10 \times 10^6)(1000 \times 10^{-12})} = 15.9 \text{ ohms}$$

$$Z_L = 50 - j15.9$$

$$Z_i = \left(\frac{N_p}{N_s}\right)^2 Z_L = \left(\frac{3}{1}\right)^2 (50 - j15.9)$$

$$Z_i = 450 - j143.1$$

Note that the capacitive reactance at the secondary in Example 8-10 was
transformed into a capacitive reactance looking into the primary. An inductive
reactance would also have been transformed into an inductive reactance.

In a non-ideal transformer with a coefficient of coupling less than 1.0, a
technique often used to transform resistive impedances from one value to another
is to resonate the primary and secondary inductances with capacitors. This technique, of course, is valid only at one frequency and the circuit is known as a double-

tuned circuit. Another technique that is sometimes used is the single-tuned circuit where the primary inductance is tuned out by selecting the proper value of capacitor in the secondary.

Transformers are used in RF circuits when isolation between sections of a circuit is desirable. In addition, transformers provide a measure of selectivity when the single-tuned or double-tuned configurations are utilized. Such configurations are equivalent to a band pass filter.

8.7 MATCHING NETWORKS

Conjugate matching networks are required in RF circuits when the complex input or output impedance of an active device such as a transistor must be matched to a load impedance, often $50 + j0$ ohms. Device impedances may be given in either series or parallel form on data sheets, and the need often arises to convert from one form to the other to utilize simple equations for matching circuit design. Figure 8-39 illustrates the conversion of a series circuit into its parallel equivalent. In Figure 8-39, if the series circuit values are known, the parallel equivalents are given by

$$R_p = R_s \left[1 + \left(\frac{X_s}{R_s} \right)^2 \right] \qquad X_p = \frac{R_p}{\left(\frac{X_s}{R_s} \right)}$$

Figure 8-40 illustrates the conversion of a parallel circuit into its series equivalent. In Figure 8-40, if the parallel circuit values are known, the series equivalents are given by

$$R_s = \frac{R_p}{1 + \left(\frac{R_p}{X_p} \right)^2} \qquad X_s = R_s \left(\frac{R_p}{X_p} \right)$$

In a typical active device at frequencies less than 1 GHz, the reactance X_p or X_s is capacitive. Figure 8-41 illustrates the two equivalent circuits for the output (or input) of an active device.

The initial step in the design of any tuned matching network is to select a value for its Q or its selectivity. A single-frequency RF circuit would normally

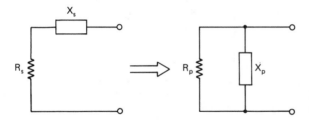

Figure 8-39 Series-to-parallel conversion.

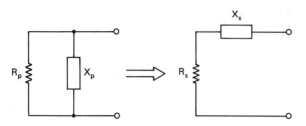

Figure 8-40   Parallel-to-series conversion.

benefit from the highest practical value of Q, while an RF circuit that must operate over some fixed bandwidth would require a value of Q consistent with this bandwidth.   The maximum Q for the matching network examples given below is approximately 10 to 20; higher values may result in unrealizable component values.

Figure 8-42 illustrates an impedance matching network, called the Pi network, which is normally used in the situation when the output resistance of the device to be matched is greater than the load resistance.   As most transistor impedance levels will be less than the commonly used load impedance of 50 ohms, this circuit is not useful for matching such devices, unless a double Pi section is used with the Q of the first section set at a very low value.

The design procedure for the Pi network is as follows:

1. Select the desired value for Q.

2. Calculate $X_{c_1} = \dfrac{R_p}{Q}$

3. Calculate $X_{c_2} = R_L \sqrt{\dfrac{\left(\dfrac{R_p}{R_L}\right)}{(Q^2 + 1) - \left(\dfrac{R_p}{R_L}\right)}}$

4. Calculate $X_L = \dfrac{QR_p + \left[\dfrac{\left[\left(\dfrac{R_p}{R_L}\right)\right]}{X_{c_2}}\right]}{Q^2 + 1}$

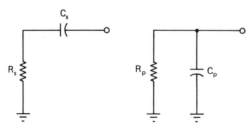

Series representation   Parallel representation

Figure 8-41   Series and parallel representation of active device input or output impedances.

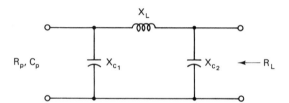

Figure 8-42 Pi network.

5. Convert the reactance into inductance and capacitance values at the operating frequency.

Note that device impedances must be in parallel form to utilize these equations.

Example 8-11

Design a Pi network to match an active device with an output impedance of 1000 ohms in parallel with 100 pf at 30 MHz to a 50 ohm load. The network Q is to be 10.

Solution

1. $Q = 10$

2. $X_{c_1} = \dfrac{R_p}{Q} = \dfrac{1000}{10} = 100 \ \Omega$

3. $X_{c_2} = R_L \sqrt{\dfrac{\left(\dfrac{R_p}{R_L}\right)}{(Q^2 + 1) - \left(\dfrac{R_p}{R_L}\right)}} = 50 \sqrt{\dfrac{20}{101 - 20}} = 24.8 \ \Omega$

4. $X_L = \dfrac{QR_p + \left[\dfrac{\left(\dfrac{R_p}{R_L}\right)}{X_{c_2}}\right]}{Q^2 + 1} = \dfrac{(10)(1000) + \dfrac{20}{24.8}}{101} = 99 \ \Omega$

5. $C = \dfrac{1}{2\pi f X_c} \qquad L = \dfrac{X_L}{2\pi f}$

$C_1 = \dfrac{1}{(6.28)(30 \times 10^6)(100)} = 53.1 \ \text{pF}$

$C_2 = \dfrac{1}{(6.28)(30 \times 10^6)(24.8)} = 214 \ \text{pF}$

$L = \dfrac{99}{(6.28)(30 \times 10^6)} = 0.53 \ \mu\text{H}$

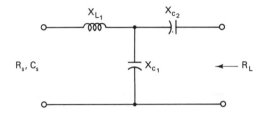

Figure 8-43   Matching network for $R_s < R_L$.

Standard capacitor values of $C_1 = 51$ pF and $C_2 = 220$ pF would be chosen for this network.

Figure 8-43 illustrates a matching network that is useful when the output or input resistance of the device to be matched is less than the load resistance, which is the normal situation in transistor amplifiers. The design procedure for the network in Figure 8-43 is as follows:

1. Select the desired value for Q
2. Calculate $X_{L_1} = QR_s + X_{c_s}$

3. Calculate $A = \sqrt{\left[\dfrac{R_s(1 + Q^2)}{R_L}\right] - 1}$

4. Calculate $B = R_s (1 + Q^2)$
5. Calculate $X_{c_2} = AR_L$

6. Calculate $X_{c_1} = \dfrac{B}{Q - A}$

7. Convert the reactances into inductance and capacitance values at the operating frequency.

Example 8-12

Design a matching network at 50 MHz to match an active device with an output impedance of $5 - j2$ ohms to a $50 + j0$ ohm load impedance. The network Q is to be 5.

Solution

(Use the network in Figure 8-43.)

1. $Q = 5$
2. $X_{L_1} = QR_s + X_{c_s} = (5)(5) + 2 = 27 \ \Omega$

3. $A = \sqrt{\left[\dfrac{R_s(1 + Q^2)}{R_L}\right] - 1} = \sqrt{\dfrac{5(26)}{50} - 1} = 1.26$

4. $B = R_s (1 + Q^2) = 5(26) = 130$

5. $X_{c_2} = AR_L = (1.26)(50) = 63\ \Omega$

6. $X_{c_1} = \dfrac{B}{Q - A} = \dfrac{130}{5 - 1.26} = 34.8\ \Omega$

7. $L_1 = \dfrac{X_{L_1}}{2\pi f} = \dfrac{27}{(6.28)(50 \times 10^6)} = 86\ \text{nH}$

   $C_1 = \dfrac{1}{2\pi f X_{c_1}} = \dfrac{1}{(6.28)(50 \times 10^6)(34.8)} = 91.5\ \text{pF}$

   $C_2 = \dfrac{1}{2\pi f X_{c_2}} = \dfrac{1}{(6.28)(50 \times 10^6)(63)} = 50.6\ \text{pF}$

Standard values of $C_1 = 91$ pF and $C_2 = 51$ pF would be chosen for this network.

Figure 8-44 illustrates a network known as the T network which is useful for matching devices with an output or input resistance less than or greater than the load resistance.   The design procedure for the T network in Figure 8-44 is as follows:

 1. Select the desired value for Q
 2. Calculate $A - R_s (1 + Q^2)$
 3. Calculate $B = \sqrt{\left(\dfrac{A}{R_L}\right) - 1}$
 4. Calculate $X_{L_1} = R_s Q + X_{c_s}$
 5. Calculate $X_{L_2} = R_L B$
 6. Calculate $X_{c_1} = \dfrac{A}{Q + B}$

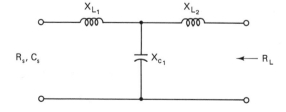

Figure 8-44   T network.

7. Convert the reactances into inductance and capacitance values at the operating frequency.

Example 8-13
Design a T network to realize the desired match in Example 8-12.

Solution

1. $Q = 5$
2. $A = 5(1 + 25) = 130$

3. $B = \sqrt{\left(\dfrac{A}{R_L}\right) - 1} = \sqrt{\dfrac{130}{50} - 1} = 1.26$

4. $X_{L_1} = R_s Q + X_{c_s} = (5)(5) + 2 = 27 \ \Omega$

5. $X_{L_2} = R_L B = (50)(1.26) = 63 \ \Omega$

6. $X_{c_1} = \dfrac{A}{Q + B} = \dfrac{130}{5 + 1.26} = 20.8 \ \Omega$

7. $L_1 = \dfrac{X_{L_1}}{2\pi f} = \dfrac{27}{(6.28)(50 \times 10^6)} = 86 \text{ nH}$

   $L_2 = \dfrac{X_{L_2}}{2\pi f} = \dfrac{63}{(6.28)(50 \times 10^6)} = 201 \text{ nH}$

   $C_1 = \dfrac{1}{2\pi f X_{c_1}} = \dfrac{1}{(6.28)(50 \times 10^6)(20.8)} = 153 \text{ pF}$

The standard value of $C_1 = 150$ pF would be chosen for this network.

The network in Figure 8-45 is an alternate configuration used when $R_s < R_L$. The design procedure for the network in Figure 8-45 is as follows:

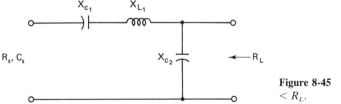

R_s, C_s

Figure 8-45   Matching network for $R_s < R_L$.

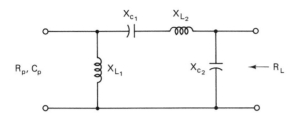

Figure 8-46 Matching network for R_s < R_L.

1. Select the desired value for Q
2. Calculate $X_{c_1} = QR_s$

3. Calculate $X_{c_2} = R_L \sqrt{\dfrac{R_s}{R_L - R_s}}$

4. Calculate $X_{L_1} = X_{c_1} + \dfrac{R_s R_L}{X_{c_2}} + X_{c_s}$

5. Convert the reactances into inductance and capacitance values at the operating frequency.

Figure 8-46 illustrates a network configuration which can also be used when R_s < R_L. Note in Figure 8-46 that the device impedances must be in parallel form. The design procedure for the network in Figure 8-46 is as follows:

1. Select the desired value for Q
2. Calculate $X_{L_1} = X_{c_p}$
3. Calculate $X_{c_1} = QR_p$

4. Calculate $X_{c_2} = R_L \sqrt{\dfrac{R_p}{R_L - R_p}}$

5. Calculate $X_{L_2} = X_{c_1} + \left(\dfrac{R_p R_L}{X_{c_2}} \right)$

6. Convert the reactances into inductance and capacitance values at the operating frequency.

Each of the networks given in this section provides a conjugate match to the device (for maximum power transfer) when terminated with the load resistance used in the design computations. However, in some applications, such as low noise RF amplifiers, a conjugate match is not necessarily appropriate, and more involved network calculations are necessary to optimize the design.

EXERCISES ━━

1. Design a 5-element capacitive input Butterworth low pass filter with a cutoff frequency of 54 MHz and an impedance of 50 ohms.
2. Redesign the filter from Exercise 1 for an impedance of 93 ohms.
3. Design a 5-element inductive input Butterworth low pass filter with a cutoff frequency of 54 MHz and an impedance of 50 ohms.
4. Design a 50-ohm, 7-element capacitive input Chebyshev low pass filter with a cutoff frequency of 120 MHz.
5. Design a 50-ohm, fifth-degree, elliptic low pass filter with a cutoff frequency of 30 MHz.
6. Design a 50-ohm, 10-element Butterworth band pass filter with a pass band of 50–54 MHz.
7. Design a 50-ohm, 5-element Chebyshev high pass filter with a cutoff frequency of 88 MHz.
8. Design a 50-ohm, 10-element Butterworth band reject filter with a center frequency of 40 MHz and a bandwidth of 6 MHz.
9. Design an inverting second order Butterworth low pass active filter with a gain of 10 and a cutoff frequency of 50 KHz.
10. Design a non-inverting second order Chebyshev low pass active filter with a gain of 8, a cutoff frequency of 20 KHz, and a ripple width of 0.1 dB.
11. Design an inverting second order Chebyshev high pass active filter with a gain of 5, a cutoff frequency of 14 KHz, and a pass band ripple width of 0.1 dB.
12. Design an inverting second order band pass active filter with a center frequency of 25 KHz, a bandwidth of 5 KHz, and a gain of 6.
13. A 50 MHz RF signal is applied to an ideal transformer whose secondary is terminated with an impedance of $50 + j30$. The turns ratio of the transformer is 5:1. Calculate the impedance looking into the input of the transformer.
14. Design a Pi network to match an active device with an output impedance of 780 ohms in parallel with 48 pF (at 50 MHz) to a 50-ohm load. The Q of the network is to be 10.
15. Design a matching network to match an active device with an output impedance of $10 - j6$ ohms to a 50-ohm load at 120 MHz. The Q of the network is to be 10.

9

Signal Processing and Conversion Circuits

Chapter 9 will discuss a number of important communications circuits under the general heading of signal processing and conversion circuits, and which operate on low level communications signals to provide some sort of transformation. The circuits to be examined include phase locked loops, companders, analog-to-digital and digital-to-analog converters, and codecs.

The phase locked loop (PLL) is one of the most widely used and versatile circuits available to the communications system designer. Phase locked loops find application as FM demodulators, FSK demodulators, signal conditioners, elements in digital frequency synthesizers, data synchronizers, AM demodulators, tone detectors, and in stereo decoders. They are also applied in areas outside of communications, such as in the motor speed controls in some tape recorder, optical disk, and computer disk drives. The linear phase locked loop is the most often used PLL type. The linear PLL is available from a number of manufacturers in integrated circuit form and implements its functions entirely with analog building blocks. The term digital phase locked loop may apply to a class of PLLs with a mix of digital and analog functional blocks, but with a digital input and output; it may also apply to a phase locked loop implemented entirely in digital hardware; or it may apply to a PLL implemented in software on a microprocessor. Digital phase locked loops are also available in integrated circuit form.

Companders are a class of integrated circuit used to compress and expand an analog waveform. Companders are used in analog communications systems, such as amplitude compandered single sideband systems, in digital communications prior to analog-to-digital conversion in PCM circuits, and in codecs.

Analog-to-digital and digital-to-analog converters are the primary links between a largely analog world and digital communications systems. Like the phase locked loop, A/D and D/A converters find numerous applications outside of the field of communications. These circuits, for example, find wide application in microprocessor-based control systems.

Codecs (derived from the two words, *co*der and *dec*oder) are integrated circuits used primarily in the telecommunications industry for A/D and D/A conversion of an analog signal, such as voice, on a single chip.

9.1 LINEAR PHASE LOCKED LOOPS

A linear phase locked loop is a feedback control system consisting of a phase comparator, a low pass filter, an error amplifier, and a voltage controlled oscillator. The block diagram of a linear phase locked loop is illustrated in Figure 9-1. In Figure 9-1, the phase comparator is a circuit that provides an output (or error) signal, $v_e(t)$, which is proportional to the phase difference between its two input signals, $v_i(t)$ and $v_o(t)$. The loop filter is a low pass filter whose bandwidth partially determines certain parameters of the phase locked loop, such as its capture range. The error amplifier controls the loop gain, which determines the phase locked loop parameter known as the lock range. The voltage controlled oscillator (VCO) is

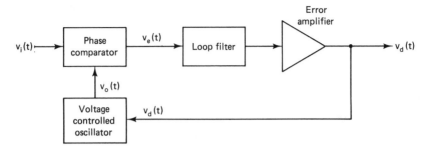

Figure 9-1   Linear phase locked loop block diagram.

a circuit that oscillates at a free running frequency f_o when its control voltage v_d is zero.   If the VCO control voltage is increased, the frequency of the output signal, $v_o(t)$, will increase; if the control voltage is decreased, the frequency of the output signal will decrease.

To understand the operation of the phase locked loop, assume that the frequency of the input signal, $v_i(t)$, is equal to the VCO free running frequency f_o. As long as this situation holds, the output of the phase comparator is zero, and the input to the VCO, $v_d(t)$, is also zero.   If it is now assumed that the input signal changes rapidly by an amount Δf, the phase of the VCO output signal, $v_o(t)$, will begin to lag the phase of the input signal.   This situation causes the phase detector to produce an error signal $v_e(t)$.   After a delay introduced by the loop filter, the control voltage to the VCO, $v_d(t)$, will increase.   This increase in control voltage causes the output frequency of the VCO to increase, which reduces the phase error to zero or to a very small value depending upon the type of loop filter used.   The VCO oscillates at the new input frequency until the input frequency changes again. The phase locked loop tracks the input signal in frequency and thus is said to be locked onto it.

Two of the most important phase locked loop parameters are its lock range and its capture range.   Lock range is the range of frequencies in the vicinity of the free running frequency f_o over which the PLL can maintain frequency lock with an input signal.   The lock range is sometimes known as the tracking range or holding range.   The lock range is increased as loop gain is increased.   The capture range is the band of frequencies in the vicinity of f_o where the PLL can acquire lock with an input signal and is sometimes called the acquisition range. The capture range is always smaller than the lock range and decreases as the loop's filter bandwidth is decreased.

The primary factors contributing to phase locked loop performance are the loop's gain and its filter characteristics.   Loop gain is defined as:

$$K_v = K_d K_o K_f A$$

In this expression, K_d is the transfer function of the phase comparator expressed in volts per radian, K_o is the transfer function of the voltage controlled oscillator expressed in radians per second per volt, K_f is the DC gain of the low pass filter,

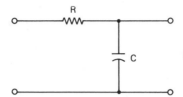

Figure 9-2 First order low pass loop filter

and A is the gain of the error amplifier. The most common loop filter is the first order low pass filter illustrated in Figure 9-2. The time constant of the simple filter in Figure 9-2 is:

$$\tau = RC$$

A first order loop filter, such as the one in Figure 9-2, results in a second order phase lock loop. A second order filter would result in a third order phase lock loop. Second and higher order filters are used in a PLL only when specific responses are desired. Third order and higher phase locked loops must be designed using control system stability analysis to avoid oscillations and/or undesirable transient response.

The lock range of a phase locked loop is given by

$$2\omega_L = 2K_v$$

$2\omega_L$ is the two-sided lock range of the PLL. The capture range of a phase locked loop with a first order loop filter is given by

$$2\omega_c = 2\sqrt{\frac{K_v}{\tau}}$$

In this expression $2\omega_c$ is the two sided capture range, and τ is the time constant of the loop filter. This expression is valid if $\tau >> 1/(2\omega_L) >> 1/(2K_v)$. The transient response of a second order phase locked loop is described by its undamped natural frequency, ω_n and its damping ratio, ζ. The undamped natural frequency of a second order PLL is

$$\omega_n = \sqrt{\frac{K_v}{\tau}}$$

The damping factor for this second order PLL is

$$\zeta = \frac{1}{2\sqrt{K_v\tau}} = \frac{\omega_n}{2K_v}$$

An often desired parameter is the capture time t_c. Capture time is the time required for the PLL to lock onto a signal after it falls within its capture range. Capture time is

$$t_c \approx \frac{1}{\omega_n}$$

The transient response of a control system, such as this second order PLL, describes how the system responds to a step change in its input. A system is said to be underdamped if $\zeta < 1$, critically damped if $\zeta = 1$, and overdamped if $\zeta > 1$. An underdamped system will respond to a step change in its input frequency by a VCO output frequency change that rises rapidly, overshoots its final value and oscillates about this final value with frequency ω_n. The oscillations decay at a rate dependent upon the value of the damping ratio until the final value is reached. In an overdamped system with $\zeta > 1$, the final value is approached slowly without any overshoot. Critical damping with $\zeta = 1$, results in the maximum rate of increase toward the final value with no overshoot. $\zeta = 0.7$ is often considered a desirable damping ratio.

An example of an integrated circuit phase locked loop is the Exar XR-S200 multi-function PLL System. The XR-S200 contains an analog multiplier that can be used as a phase detector, a voltage controlled oscillator, and an operational amplifier. Each of these functions may be utilized independently or together with an external loop filter to implement a PLL system. The XR-S200 is usable from 0.1 Hz to 30 MHz and can be powered by supplies ranging from ± 3 volts to ± 30 volts. Another example of an integrated circuit PLL is the Signetics NE565, which is a general-purpose PLL designed to operate at frequencies between 0.001 Hz and 500 KHz. The NE565 operates from ± 6 volts to ± 12 volts. Both the XR-S200 and the NE565 allow external programming of the free running VCO frequency, and in addition, the XR-S200 allows digital programming of f_o from $1.0f_o$ to $2.5f_o$ in $0.5f_o$ steps. The XR-S200 also has provisions for frequency sweeping, on-off keying and synchronization of the VCO to an external sync pulse.

Example 9-1

A phase locked loop system is implemented with a phase detector that has a transfer function of 0.1 volts per radian, a VCO with a transfer function of 10^6 radians per second per volt, an operational amplifier with a gain of 5 and a first order loop filter with $R = 1$ K and $C = 0.01$ μF. Compute the lock range, the capture range, the undamped natural frequency, the damping factor and the capture time for this PLL.

Solution

$$K_v = K_d K_o K_f A = (0.1)(10^6)(1)(5) = 5 \times 10^5$$

$$\tau = RC = (10^3)(0.01 \times 10^{-6}) = 1 \times 10^{-5} \text{ sec}$$

$$2f_L = \frac{2\omega_L}{2\pi} = \frac{2K_v}{2\pi} = \frac{5 \times 10^5}{3.14} = 159 \text{ KHz}$$

$$2f_c = \frac{2\omega_c}{2\pi} = \frac{2\sqrt{\dfrac{K_v}{\tau}}}{2\pi} = \frac{2\sqrt{\dfrac{5 \times 10^5}{1 \times 10^{-5}}}}{2(3.14)} = 71.2 \text{ KHz}$$

Note that $\tau \gg \dfrac{1}{2K_v}$ or $1 \times 10^{-5} \gg \dfrac{1}{(2)(5 \times 10^5)} \gg 10^{-6}$.

$$f_n = \frac{\omega_n}{2\pi} = \frac{\sqrt{\dfrac{K_v}{\tau}}}{2\pi} = \frac{\sqrt{\dfrac{5 \times 10^5}{1 \times 10^{-5}}}}{6.28} = 35.6 \text{ KHz}$$

$$\zeta = \frac{\omega_n}{2K_v} = \frac{2\pi f_n}{2K_v} = \frac{(6.28)(35.6 \times 10^3)}{(2)(5 \times 10^5)} = 0.22$$

$$t_c \approx \frac{1}{\omega_n} = \frac{1}{2\pi f_n} = \frac{1}{(6.28)(35.6 \times 10^3)} = 4.5 \text{ }\mu\text{sec}$$

This system is highly underdamped and will respond quickly to a step change in input frequency, but will oscillate about the final value for a considerable amount of time. The damping ratio can be increased by decreasing the loop gain and/or loop filter time constant.

9.2 DIGITAL PHASE LOCKED LOOPS

A digital phase locked loop (DPLL) may be implemented with a digital phase comparator, an analog loop filter, an analog error amplifier, and an analog voltage controlled oscillator whose output is a square wave. Figure 9-3 illustrates the most widely used digital phase comparator circuitry. The outputs of the phase comparator in Figure 9-3 are used to control a circuit, often called a charge pump, that provides an analog signal to the loop filter. The charge pump works as follows: if the UP output is active low, a fixed positive voltage is applied to the filter input; when the DOWN output is active (low), a negative voltage (of the same magnitude as the positive voltage) is applied to the filter input. When both the UP and DOWN outputs are high, no voltage is applied to the filter input. The phase comparator then works as follows: if the output signal $v_o(t)$ lags the input signal $v_i(t)$, the UP output of the phase comparator will generate pulses with a duty cycle proportional to the phase error and the DOWN output will be inactive high; if the input signal $v_i(t)$ lags the output signal $v_o(t)$, the DOWN output will generate pulses and the UP output will remain high; if the two signals are in phase, both UP and DOWN will remain high. The charge pump converts the pulses to an analog signal, which is applied to the loop filter. This type of phase comparator is also sensitive to frequency offsets, which accounts for its popularity.

The digital phase locked loop implemented with a phase comparator, such as the one illustrated in Figure 9-3, has, at least theoretically, infinite capture range and lock range. The transient response of this digital PLL is identical to that of the linear PLL.

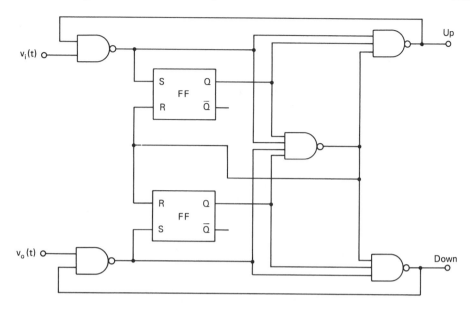

Figure 9-3   Digital phase comparator circuitry.

A second type of digital phase locked loop, consisting entirely of digital hardware elements, may be implemented with a digital phase comparator, a digital loop filter and a digitally controlled oscillator (DCO). The error amplifier, an analog element, is of course absent from this PLL implementation. The input signal to a digital hardware phase locked loop may be digital or, if the DPLL includes an A/D converter, analog.

The digital phase comparator was described previously. A type of phase comparator with one analog input, an N-bit digital word input, and an N-bit digital word output is the Nyquist rate phase comparator illustrated in Figure 9-4. The digital multiplier operates in a fashion similar to the analog multiplier used in most linear phase detectors. In a digital hardware PLL, however, the multiplier is implemented with digital hardware.

The digital loop filter used in this type of PLL must be compatible with the

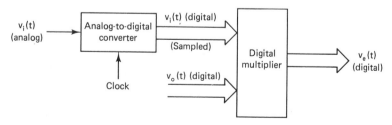

Figure 9-4   Nyquist rate phase comparator.

type of phase detector used. The output of a digital phase comparator is UP or DOWN pulses, while the output of the Nyquist rate phase comparator is an N-bit digital word. In the type of PLL discussed in this section, the digital filter is implemented entirely in hardware.

A digitally controlled oscillator is either a divide-by-N counter or a waveform synthesizer, which is controlled by an N-bit digital control input from the digital loop filter.

An example of a digital hardware phase locked loop available in integrated circuit form is the Texas Instruments SN74LS297. This circuit uses a simple exclusive or circuit as a phase comparator and a simple up-down counter for a loop filter. The DCO is also implemented with counters.

All of the functions indicated in the digital hardware phase locked loop may be implemented in software on a microprocessor. Because microprocessors are relatively slow, this implementation is limited to low frequencies and coprocessor chips are normally used. The phase detector function is either the software equivalent of a digital phase detector or a software multiply operation. As discussed in Section 8.5, the digital filter block may easily be implemented in software. The digitally controlled oscillator is also straightforward to implement in software by the use of look-up tables. The major advantage of a digital software PLL is the ability to adjust the filter performance with minor program modifications. As microprocessor performance is improved, this type of PLL will gain increased popularity.

9.3 COMPANDERS

Companders may be implemented entirely with analog functions or digitally using piecewise approximations to the desired transfer function. Digital piecewise approximations to the two companding laws used in telecommunications, the μ-law and the A-law, are used in the integrated circuit codecs (coder-decoder chips), which will be discussed later in this chapter. Analog companders utilize three primary functional blocks: a rectifier circuit, a variable gain cell, and an operational amplifier. The rectifier circuit with an associated filter capacitor provides a control voltage proportional to the average value of its input signal. The rectifier output is used to control the gain of the variable gain cell. The operational amplifier is necessary to provide additional gain and to provide an easily implemented compressor circuit.

The Signetics NE570 is an example of an integrated circuit analog compander. The NE570 has two identical channels, each of which may be configured as an expander or as a compressor. Figure 9-5 illustrates the block diagram of one of the two identical channels of the NE570. The NE570 provides a fixed 2:1 compression ratio in dB. A 2 dB level change on the input of the compressor becomes a 1 dB change at its output; the receiving expander provides a complementary expansion. Figure 9-6 illustrates the basic NE570 expander configuration.

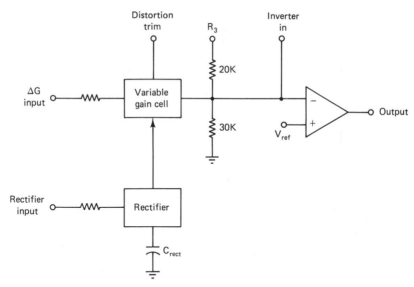

Figure 9-5   Signetics NE570 compander block diagram (courtesy of Signetics Corporation).

Figure 9-7 illustrates the NE570 in its compressor configuration.   In Figure 9-7, the R_{dc} resistors provide a DC feedback path and determine the bias level at the output of the operational amplifier.   The variable gain cell and the rectifier provide a feedback path at AC only, which implements the compression function.

Companders can provide significant improvements in signal-to-noise ratios and are used in high fidelity applications, in high-performance automatic level

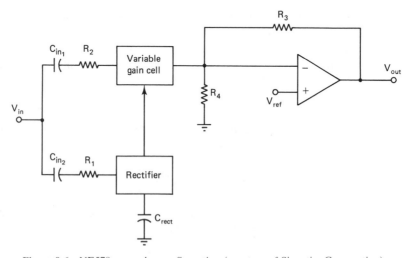

Figure 9-6   NE570 expander configuration (courtesy of Signetics Corporation).

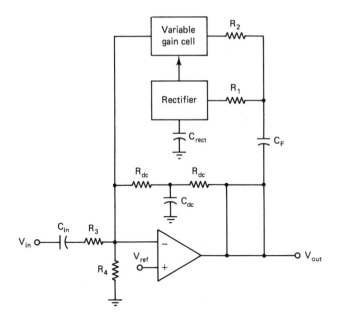

Figure 9-7 NE570 compressor configuration (courtesy of Signetics Corporation).

control systems, and as voltage controlled attenuators in addition to the previously mentioned applications in ACSB and PCM. The μ-law and A-law companders used in telecommunications applications are normally integrated directly on codec chips.

9.4 DIGITAL-TO-ANALOG CONVERTERS

The digital-to-analog (D/A) converter is an essential element in a digital communications system that must provide an analog output, such as voice or video. It may be thought of as a device that accepts a digital word D_i and provides an analog output voltage V_o which is related to the digital input by the following equation:

$$V_o = \alpha D_i$$

In this expression α is a reference voltage; D_i is the digital input word which may be expressed as

$$D_i = \frac{b_1}{2^1} + \frac{b_2}{2^2} + \frac{b_3}{2^3} + \ldots \frac{b_n}{2^n}$$

In this expression, b_1, b_2 . . . b_n are the bit values (1 or 0) of the input word. A generalized expression for a D/A converter is then

$$V_o = \alpha \, [b_1 2^{-1} + b_2 2^{-2} + b_3 2^{-3} + \ldots b_n 2^{-n}]$$

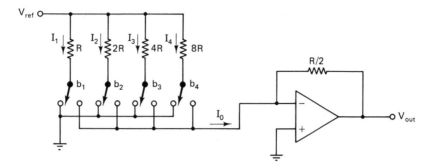

Figure 9-8   Binary weighted digital-to-analog converter.

The above equation may be implemented with the circuit illustrated in Figure 9-8, which is known as a binary weighted circuit. In Figure 9-8, the relative weights of the currents I_1, I_2, I_3 and I_4 are set by the resistor network so that the output voltage of the operational amplifier implements the D/A equation. In practice, the binary switches illustrated in Figure 9-8 are implemented electronically.

Note that the resistance values in the 4-bit D/A converter in Figure 9-8 vary from $R/2$ to $8R$, a 16:1 ratio. This resistance spread is difficult to achieve in monolithic integrated circuits without expensive selection and trimming operations. For D/A converters with more than a four-bit input, an alternative is mandatory. Figure 9-9 illustrates the so-called R-$2R$ D/A converter circuit that solves the resistance spread problem, but which requires additional resistors. Note that the circuits in Figures 9-8 and 9-9 switch from ground to the virtual ground at the input to the operational amplifier. This technique is known as current-switching and is the method most often used in integrated circuit D/A converters, because it minimizes switching transients and thus maximizes speed.

Digital communications systems normally utilize D/A converters, which are

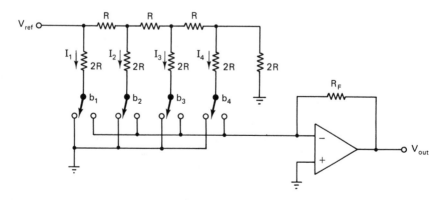

Figure 9-9   R-2R digital-to-analog converter.

available in integrated circuit form.   Key specifications to consider when selecting a D/A converter are as follows:

- Settling time—the time required for the output to settle to and remain within a specified error band about the final value.
- Resolution—determined by the number of input bits, N, of the D/A converter.   The resolution equals one part in 2^N.
- Offset error—the constant shift from the ideal transfer characteristic of the D/A converter; the output obtained with an input code corresponding to zero output.
- Differential non-linearity—the deviation of the measured output step size from the ideal step size.
- Integral non-linearity—the difference between the ideal transfer characteristic and the actual transfer characteristic.

Digital to analog converter chips are sold by virtually every semiconductor manufacturer, and hundreds of different device types are available.   D/A converters are available with 4-, 8-, 10-, 12-, 14-, or 16-bit inputs and either voltage outputs as illustrated in Figures 9-7 and 9-8 or current outputs.   Current output D/A converters often require an external operational amplifier configured as a current to voltage converter but offer increased flexibility.

One example of an integrated circuit digital-to-analog converter is the Signetics NE5410, which is a 10-bit high speed D/A converter available in a 16-pin dual in-line package.   The NE5410 operates from a 5-volt supply and has a settling time of 250 nsec.   The device has a current output that ranges from 0 mA (for all of the input bits set low) to 4 mA (for all of the input bits set high).

9.5 ANALOG-TO-DIGITAL CONVERTERS

The device at the input of a digital communications system that converts an analog signal, such as voice or video, to a digital signal is called an analog-to-digital (A/D) converter.   The A/D converter performs the reverse operation of a D/A converter—it encodes an analog voltage into a digital word of predetermined length.

The most straightforward type of analog-to-digital converter is the tracking A/D converter illustrated in Figure 9-10.   The tracking A/D converter operates as follows: assuming the counter starts at zero, it counts clock pulses.   The binary count is converted to an analog voltage by the D/A converter and compared to the analog input with a comparator.   When the two voltages are equal, the comparator changes state, which terminates the count.   The binary count at that point is the digital word corresponding to the analog input voltage.   This type of converter is very slow for full-scale changes in input voltage, since it requires 2^N-1 clock cycles to be counted.   For small changes in input voltage, the tracking D/A is fast, since only a few clock cycles need to be counted.

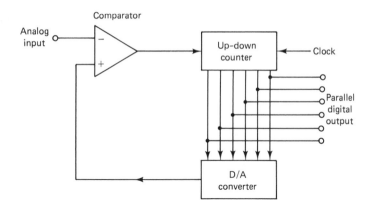

Figure 9-10 Tracking analog-to-digital converter.

The most common type of A/D converter is the successive approximation type, illustrated in Figure 9-11.   The successive approximation A/D converter works as follows: the analog input is approximated by trying a digital one in each successive bit, starting with the most significant bit.   This type of converter requires only a maximum of N clock cycles for N-bit resolution and thus is relatively fast.

Key specifications in an A/D converter include resolution, offset error, dif-

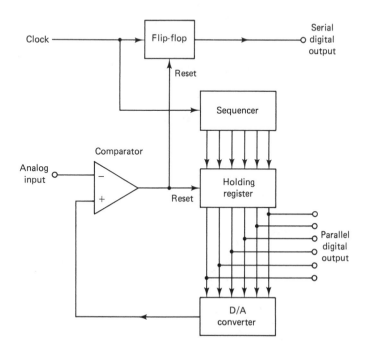

Figure 9-11 Successive approximation analog-to-digital converter.

ferential non-linearity and integral non-linearity described previously in the section on digital to analog converters, in addition to the following:

- Conversion time—time required for a complete conversion cycle in an A/D converter.
- Gain error—error in the slope of the transfer characteristic of the A/D converter.
- Quantizing error—error due to finite number of output codes; this can be as great as 1/2 of the value of the least significant bit.

An example of an analog-to-digital converter is the Signetics NE5034 8-bit A/D converter. The NE5034 uses the successive approximation technique and operates from +5 and −12 volt supplies. The device is available in an 18-pin dual in-line package. The NE5034 features a conversion time of 17 μsec using its 500 KHz internal clock and approximately 11 μsec using an external 700 KHz clock. This converter can accommodate a bipolar or unipolar input voltage.

As with digital-to-analog converters, hundreds of different types of analog-to-digital converters are available in resolutions up to 16 bits.

9.6 CODECS

A codec (*co*der-*dec*oder) is an integrated circuit with a PCM transmitter and receiver integrated onto a single chip. They are used primarily in telecommunications applications but may also be used in voice store-and-forward systems, digital echo cancellers, secure communications systems, satellite earth stations, data acquisition systems, signal processing systems, and telemetry applications.

Codecs contain a complete PCM baseband transmitter and receiver system including a compander (μ-law, A-law or selectable), an analog-to-digital converter, a digital-to-analog converter and the necessary control logic. Codecs are available with and without integrated filters. Codec filters typically utilize switched capacitor filter technology. Consistent with their primary utilization in voice telecommunications applications, the most commonly available codecs sample the analog input at an 8 KHz rate and may operate at a digital data rate of 64 Kb/s for applications where a single serial bus is dedicated to a single codec, or at a higher data rate when many codecs are time division multiplexed onto a single bus.

An example of a single chip codec with integrated filters is the Motorola MC14400. The MC14400 is available in a 16-pin dual in-line package and is fabricated using low-power complementary MOS technology (CMOS), which allows it to have flexible power supply requirements. Either μ-law or A-law companding is pin-selectable, as is the format of the digital inputs and outputs to allow the chip to conform to either of the two primary international standards.

Another example of a single chip codec with integrated filters is the Intel 2913. The 2913 is available in a 20-pin dual in-line package and operated from

±5 volts, as it is fabricated using *N*-channel HMOS technology. The 2913 also has pin-selectable μ-law or A-law companding and operates with a fixed data rate mode or a variable data rate mode from 64 Kb/s to 2.048 Mb/s. The Intel 2914 codec is a more advanced version of the 2913; it can operate with asynchronous clocks, has additional signaling capability and has a loop-back test function. The 2914 is available in a 24-pin dual in-line package or a 28-pin leaded chip carrier package.

EXERCISES ━━━━━━━━━━━━━━━━━━━━━━━━━━━━━━━━━━

1. A phase locked loop consists of a phase comparator with a transfer function of 0.2 volts/radian, a VCO with a transfer function of $4\pi \times 10^4$ radians/second/volt, an operational amplifier with a gain of 10 and a loop filter with a DC gain of 1.0. Calculate the loop gain of this phase locked loop.
2. Calculate the lock range of the phase locked loop of Exercise 1.
3. Calculate the capture range of the phase locked loop in Exercise 1 if the loop filter is first order with a time constant of 20 μsec.
4. Calculate the natural frequency and damping factor for the phase locked loop in Exercise 1.
5. Calculate the capture time for the phase locked loop in Exercise 1.
6. An 8-bit digital-to-analog converter with a reference voltage of 5 volts receives an input word of 11001101. Compute the output voltage.

10

Modulation and Demodulation Circuits

Modulation and demodulation circuits are the heart of any communications system, since they superimpose the desired information onto and extract it from a carrier. Analog modulation techniques were discussed in Chapter 3, digital modulation methods in Chapter 6. This chapter illustrates the typical electronic circuits that implement analog and digital modulation techniques. The first two sections in this chapter provide an overview of traditional AM and FM modulation and demodulation circuits. These circuits have been in use for many years and still find application in commercial broadcast transmitters and receivers. The final two sections discuss multiplier and phase locked loop modulation and demodulation circuits. The balanced modulator, a specific type of multiplier circuit, is perhaps the most fundamental communications circuit. The versatile phase locked loop finds application in numerous modulators and demodulators as well as in other communications applications. Most current communications equipment utilizes either a multiplier circuit or a phase locked loop in its modulation and/or demodulation sections.

10.1 AM MODULATION AND DEMODULATION CIRCUITS ⎯⎯⎯⎯⎯⎯⎯

Chapter 3 illustrated a technique for generating an amplitude modulated signal using a multiplier circuit. In addition to this technique, an AM signal may be generated by using a non-linear device or by the so-called direct modulation method. The multiplier method and the non-linear device methods are known as low-level modulation methods, since they are accomplished with very low power signals which are subsequently amplified by power amplifier circuits to a level suitable for transmission. The direct modulation method is called high-level modulation, because the modulating signal is normally applied to the final power amplifier stage in the transmitter.

Non-linear device modulation is accomplished by summing the modulating signal and the carrier signal, applying the sum to a non-linear device and passing the device output through a band pass filter. The non-linear device must have no greater than a second order (square-law) non-linearity and the modulating frequency must be less than one-third of the carrier frequency for this type of modulator. If the non-linear device is a half-square law device, such as a diode, 100% modulation is not possible.

In a direct modulation scheme, the output stage is often driven by a modulated low-level stage simultaneously with modulation of the collector supply of the output stage. This arrangement maximizes the efficiency of the modulated output stage. The low-level stage and the output stage are modulated by adding the modulator output to the respective collector supplies, usually through a modulation transformer.

An amplitude modulated signal may be demodulated synchronously using a multiplier circuit, or it may be demodulated using a phase locked loop. A major advantage of AM, however, is that it may also be satisfactorily demodulated with

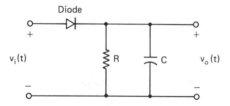

Figure 10-1 Amplitude modulation envelope detection circuit.

a simple diode and low-pass filter circuit. Figure 10-1 illustrates an amplitude modulation envelope detection circuit.

The envelope detector works by charging the capacitor to the peak value of the carrier cycle. Once the peak value has been reached, the diode becomes reverse biased, allowing the capacitor to discharge through the resistor until the next carrier cycle, when the capacitor will be charged to the next peak and so on. The time constant of the RC filter is selected to be a compromise between the amount of RF ripple and the amount of audio high frequency attenuation and distortion. An often-used design value for the filter time constant is one-tenth the period of the modulating signal.

Example 10-1

Design an envelope detector circuit for the AM broadcast band. The load presented to the preceding stage must be at least 10K ohms at DC.

Solution

The AM broadcast band in the USA runs from 540 KHz to 1600 KHz. The maximum frequency of an audio signal in the AM broadcast band is limited to 10 KHz, which determines the filter time constant. The specified maximum loading allows a resistance of 10K ohms.

$$RC = \frac{T_m}{10} = \frac{1}{10 f_m} = \frac{1}{(10)(10 \times 10^3)} = 1 \times 10^{-5} \text{ sec}$$

$$C = \frac{1 \times 10^{-5}}{R} = \frac{1 \times 10^{-5}}{10^4} = 1 \times 10^{-9} = 0.001 \ \mu\text{F}$$

The envelope detector would be the circuit illustrated in Figure 10-1 with $R = 10$K ohms and $C = 0.001 \ \mu$F. The diode selected is not critical if the input voltage to the detector is much greater than the diode turn-on voltage. At frequencies much higher than the broadcast band, the diode capacitance would be a consideration.

10.2 FM MODULATION AND DEMODULATION CIRCUITS

Frequency modulation of a carrier is normally accomplished at low levels in the transmitter circuitry prior to power amplification. One technique for generating FM using a multiplier and phase shifter was discussed in Chapter 3. Another popular technique makes use of a device known as a voltage variable capacitor (VVC). A VVC's capacitance changes as the voltage applied across its terminals changes. Voltage variable capacitors are normally implemented with reverse biased diodes. Diodes used as VVCs are called varactor diodes. If a VVC is used in the frequency-determining network in an oscillator circuit, a modulating voltage may be applied to the VVC, which causes the oscillator frequency to change as a linear function of the modulating signal to produce FM. An example of a VVC is the Motorola 1N5461A. This diode has a nominal capacitance of 6.8 pF at a reverse voltage of 4 volts and a frequency of 1 MHz. The 1N5461A has a minimum tuning ratio (2 volts to 30 volts) of 2.7 at 1 MHz and a minimum Q of 600 at 50 MHz. Motorola has a complete set of related diodes available with nominal capacitance values ranging from 6.8 pf to 100 pf. Voltage variable capacitors are also often used in the voltage controlled oscillators used in phase locked loops.

Traditional methods of demodulating FM signals include the discriminator and ratio detector. These techniques are stll in use, especially in home entertainment receivers, but are quickly being made obsolete by the use of phase locked loop FM detectors. The operation of a discriminator is illustrated by the characteristic curve in Figure 10-2.

In Figure 10-2, a frequency deviation Δf away from the carrier frequency is converted into a discriminator output voltage variation of ΔA by the sloping transfer characteristic of the discriminator. A discriminator is preceded by a circuit known as a limiter in most FM receivers. The purpose of the limiter is to remove amplitude variations due to sources such as noise and fading from the signal. A limiter is a high-gain amplifier followed by a circuit that clips the positive and negative FM signal peaks. Figure 10-3 illustrates a typical (Foster-Seely) discriminator circuit.

The discriminator circuit in Figure 10-3 operates as follows: when the input signal is constant in frequency, voltages of equal amplitude and opposite phase are induced across each half of the secondary of the transformer and are rectified and summed to give zero output. A shift in frequency causes a shift in the phase of

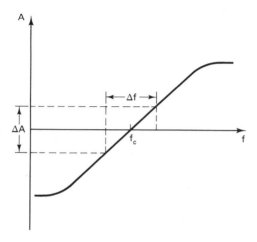

Figure 10-2 Discriminator characteristic curve.

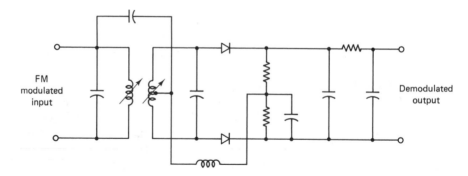

Figure 10-3 Discriminator circuit.

the secondary winding voltage, which, when added to the voltage at the center tap, causes more current to flow through one diode than the other. This difference results in a discriminator output proportional to the shift in frequency.

The ratio detector circuit eliminates the need for limiting prior to detection in non-critical applications, such as home entertainment and public service band FM receivers. The ratio detector illustrated in Figure 10-4 allows the input amplitude to vary over a large range with minimal effect on the output amplitude.

The discriminator circuit in Figure 10-3 and the ratio detector circuit in Figure 10-4 require considerable alignment effort for satisfactory performance. One solution to the alignment problem is to replace the secondary of the transformer in a discriminator with a quartz crystal in parallel with an inductor. The so-called crystal discriminator is an adjustment-free FM detector. Despite this improvement, most current designs utilize phase locked loops for FM demodulation.

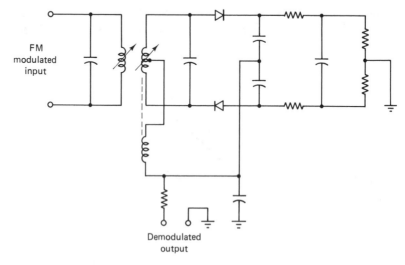

Figure 10-4 Ratio detector circuit.

10.3 MULTIPLIER MODULATION AND DEMODULATION CIRCUITS ━━━━━━

The multiplier is one of the most fundamental and important circuits in communications. Multipliers are used in numerous modulation and demodulation schemes in both analog and digital communications systems. Multipliers that operate on analog communications signals are called analog multipliers; they, however, are used in many digital communications systems, as the reader will recall from Chapter 6. The multiplication function can, of course, be implemented digitally in software, but speed limitations presently prevent widespread use of this technique at RF frequencies.

The most general multiplier circuit is the four-quadrant multiplier, which produces an output proportional to the product of its two inputs and maintains all polarity relationships. Four-quadrant multiplier circuits are available in integrated circuit form, and most depend upon the *variable transconductance* property of transistor pairs. Variations in transconductance result from variations in transistor emitter current. In a simplified multiplier circuit, one signal input is the input to a differential amplifier pair, while the other controls an emitter current source resulting in an output voltage proportional to the product of the two input voltages. Practical integrated circuit multipliers use more complex circuitry to increase the dynamic range and reduce common mode voltage errors. Four-quadrant multipliers are commonly specified by their accuracy, linearity, and bandwidth. *Accuracy* is the maximum deviation of the actual output level from the ideal level for any choice of input values within the dynamic range of the multiplier. *Linearity* is the maximum percentage deviation from a straight line at the output for equal inputs. The *bandwidth* of a multiplier describes its high frequency capability. A four-quadrant multiplier is called a balanced circuit, since by definition its output is zero if either of its inputs is zero. Four-quadrant multipliers are used as phase comparators in phase locked loops.

An example of a four-quadrant multiplier integrated circuit form is the Motorola MC1495L. The MC1495L is available in a 14-pin dual in-line package and operates from ±15 volts. The MC1495L has an adjustable scale factor with linearity of 1% (maximum error) on the X-input and 2% on the Y-input. The bandwidth of the circuit is 80 MHz with a 50-ohm load.

In most communications applications, the four-quadrant capability of the multiplier circuit is not necessary. In modulator and demodulator applications, one input of the multiplier is a constant amplitude carrier. A linear transfer characteristic is not required on this input; in fact, the amplitude of the carrier is often increased until this input saturates, which results in a rectangular waveform. The other (modulating signal) input in such a circuit must maintain a linear transfer characteristic. A multiplier circuit operated in this fashion is one form of balanced modulator. Balanced modulators allow simpler circuitry than four-quadrant multipliers and are available in integrated circuit form or may be fabricated from discrete components.

Figure 10-5 illustrates a balanced modulator circuit fabricated with two trans-

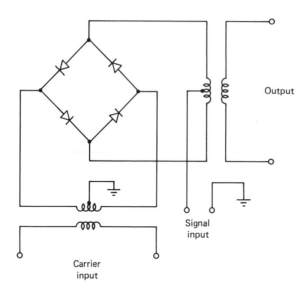

Output

Signal
input

Carrier
input

Figure 10-5 Diode ring double-balanced modulator circuit.

formers and four diodes in a ring configuration. This circuit is called a double-
balanced modulator to distinguish it from an inferior single-balanced design that
uses only two diodes.

Hot carrier diodes are normally used in balanced modulators because of their
low noise, low forward resistance, high reverse resistance, fast switching time, and
good temperature stability. An example of a hot carrier diode useful in balanced
modulator applications is the Hewlett-Packard HPA-5082-5826. Diode rings are
also available in integrated circuit form. The advantage of such circuits is the
excellent characteristic matching, which is possible when the four diodes are fabricated on a single substrate. A typical diode ring IC is the RCA CA3039. Diode
ring balanced modulators are also available with all of the components pre-packaged
in a case with coaxial connectors.

Integrated circuit balanced modulators are used when elimination of the transformers in a discrete balanced modulator and/or some gain is desirable. Integrated
circuit balanced modulators also require less carrier power than do the discrete
circuits. The major disadvantage of IC balanced modulators is their limited bandwidth. Integrated circuit balanced modulators normally utilize a differential amplifier configuration to achieve the balanced behavior.

An example of an integrated circuit balanced modulator is the Plessey SL640C. This
circuit operates from a single supply of +6 volts and has a bandwidth of 75 MHz.
The Motorola MC1496 is available in a 14-pin dual in-line package and operates
from either a single supply of +12 volts or from a dual supply. The MC1496 has
a bandwidth of 10 MHz and has a provision for external gain adjustment. Typical
carrier suppression for the MC1496 at its upper limit of 10 MHz is 50 dB.

The output spectrum of a balanced modulator consists of sidebands at $f_c - f_s$

and $f_c + f_s$, where f_c is the carrier frequency and f_s is the signal frequency. Other components are suppressed at f_c, $f_c \pm nf_s$, nf_c and $nf_c \pm nf_s$, where n is an integer. Undesirable components in the balanced modulator output may be further suppressed by filtering.

10.4 PHASE LOCKED LOOP MODULATION AND DEMODULATION CIRCUITS

The phase locked loop or some of its functional blocks may be used to realize AM and FM modulators and demodulators as well as FSK and PSK modulator and demodulator circuits. Phase locked loop circuits are widely used in communications due to their availability in integrated circuit form, low cost, and excellent performance.

A low-level AM modulator may be realized using the voltage controlled oscillator and phase detector from an integrated circuit PLL. In Figure 10-6, which illustrates the circuit, the phase detector (an analog multiplier) forms the product of the modulating signal $m(t)$ and a carrier signal $v_c(t)$, produced by the voltage controlled oscillator. If $m(t)$ is biased with a DC level, the output signal $v_o(t)$ will be an AM signal. If the modulating signal is biased at zero volts, the output will be a double sideband suppressed carrier signal. If the DSB signal is appropriately filtered, an SSB signal results.

The phase locked loop may also be used to implement a synchronous AM demodulator as illustrated in Figure 10-7. In the circuit illustrated in Figure 10-7, the PLL is used to reconstruct the carrier required for synchronous demodulation. When a phase locked loop is inlock, its VCO output is offset from the original carrier by 90°. A number of VCO circuits, however, provide a quadrature output that provides the necessary in-phase signal. The multiplier in Figure 10-7 could be a PLL phase comparator, if desired.

The voltage controlled oscillator in a PLL may be used as an FM modulator and the phase comparator as a buffer amplifier, as illustrated in Figure 10-8. In Figure 10-8, the phase comparator, which is an analog multiplier, will act as a buffer amplifier if one of its inputs is held at a constant DC level V_{DC}. Applying

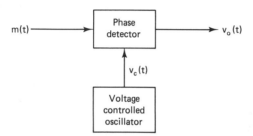

Figure 10-6   Phase locked loop component AM modulator.

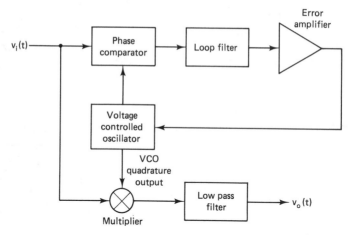

Figure 10-7   Phase locked loop AM demodulator.

a DC bias to one input of an analog multiplier is equivalent to multiplication by a constant, which is the function of an amplifier.

The basic phase locked loop circuit, as illustrated in Figure 10-9, serves well as an FM demodulator. In Figure 10-9, as the input signal changes in frequency with modulation, the PLL will attempt to track it, generating an error signal, which is the desired demodulated output.

FSK modulation and demodulation may be accomplished in a fashion very similar to FM. The FSK modulator can be implemented, as in Figure 10-8, with the addition of an amplifier between the binary modulating signal and the VCO input, so that the level of the modulating signal can be adjusted to produce the desired frequency shift in the VCO. Some PLLs designed for use as FSK modulators allow the deviation to be set with external resistors, eliminating the need for an amplifier. An FSK signal may be demodulated, as in Figure 10-9, with the addition of a comparator and logic driver circuit at the PLL output. An example of a phase locked loop circuit designed for use as an FSK demodulator is the EXAR XR-210 FSK Modulator/Demodulator. The XR-210 is available in a 16-pin dual in-line package and operates from a single supply of 5 volts to 26 volts or a split

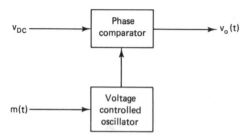

Figure 10-8   Phase locked loop component FM modulator.

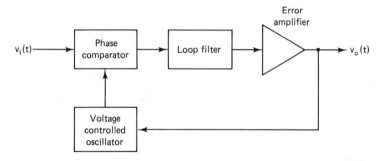

Figure 10-9   Phase locked loop FM demodulator.

supply of ± 2.5 volts to ± 13 volts; it operates from 0.5 Hz to 20 MHz. This integrated circuit includes a comparator and logic driver on the same chip as the PLL. The output can be made compatible with RS-232C levels, ECL levels, or TTL levels. The mark (logical one) and space (logical zero) frequencies may be adjusted independently in the XR-210.

The voltage controlled oscillator of a PLL and its phase comparator may be used to implement a PSK modulator, as illustrated in Figure 10-10. In Figure 10-10, the phase comparator, which is an analog multiplier, is used as a modulator. The bipolar modulating signal $m(t)$ switches the polarity of the carrier frequency at the output, which is equivalent to a 180° phase shift. Figure 10-11 illustrates a PSK receiver implemented with a phase locked loop, an absolute value circuit, a Schmitt trigger and a JK flip-flop.

When the polarity of the input signal in Figure 10-11 is switched, the PLL sees a momentary phase error of 180°, which will produce a positive or a negative error pulse depending upon the data transition. The absolute value and Schmitt trigger circuits shape the pulses, which are used to toggle a JK flip-flop. The flip-flop changes state for each transition of the error signal, which produces the original modulating signal.

In Chapters 3 and 6, many of the demodulation schemes discussed required a signal at the same frequency and in phase with the transmitted carrier. As illustrated in this section, the PLL makes an excellent carrier recovery circuit and is widely used as such.

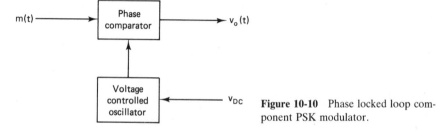

Figure 10-10   Phase locked loop component PSK modulator.

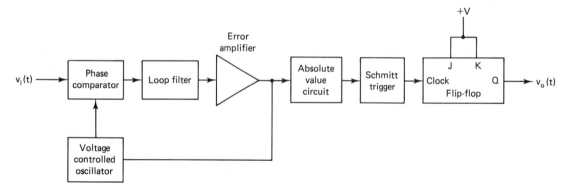

Figure 10-11   Phase locked loop PSK demodulator.

EXERCISES ━━━━━━━━━━━━━━━━━━━━━━━━━━━━━━━━

1. Design an envelope detector for a 15 MHz AM signal if the maximum modulation frequency is 5 KHz and maximum loading on the previous stage is 15K ohms.

2. A varactor diode has a tuning ratio (1 volt to 20 volts) of 3.0. The capacitance at 1 MHz and 1 volt is 10 pF. Estimate the capacitance at 15 volts using a linear approximation to the tuning curve.

3. A discriminator circuit has a characteristic curve with a slope of 4×10^{-5} volts/Hz. Compute the amplitude change for a frequency deviation of 5 KHz.

4. A 30 MHz carrier signal and a 20 KHz modulating signal are applied to a balanced modulator. Compute the unsuppressed signal frequencies at the output of the device.

5. A DC voltage of ± 5 volts is applied to one input of a phase comparator block in a phase locked loop and a 10 MHz carrier with a peak amplitude of 2 volts is applied to the other input. Calculate the output signal of the phase comparator.

11

RF Fundamentals

Most analog or digital communications systems operate at frequencies above base-band. These systems may utilize RF carrier frequencies from approximately 1 MHz to over 100 GHz. In addition, the IF or intermediate frequencies used in some superheterodyne transmitter and receiver systems range up to 70 MHz or higher. Most newer communications systems operate with carrier frequencies in the very high frequency (VHF), ultra high frequency (UHF) or microwave portions of the spectrum. The VHF/UHF range extends from approximately 30 MHz to 1000 MHz and the microwave range covers approximately 1 GHz (1000 MHz) to 30 GHz. RF frequencies above approximately 30 GHz are called the millimeter frequencies. The range of frequencies from approximately 1 to 30 MHz is called the high frequency (HF) band.

Communications circuits that operate efficiently at RF frequencies require special design techniques. The purpose of this chapter is to provide the background fundamentals necessary to understand the RF and IF portions of receivers and transmitters, transmission lines, and antennas. Some of the concepts presented in this chapter have important applications outside of the field of communications. An example would be the transmission line and microstrip concepts and techniques, which are used in high speed digital logic design.

11.1 TRANSMISSION LINES

A transmission line transfers electromagnetic energy between points in a circuit or system. At low frequencies, transmission lines are simply interconnecting pairs of wires. As the frequency is increased and transmission line dimensions approach the wavelength of the energy to be transferred, simple wire connections are no longer sufficient for efficient energy transfer, and other types of transmission lines must be used. Examples of RF transmission lines include flexible coaxial cable, semi-rigid coaxial cable, waveguide, and microstrip. At optical frequencies, fiber optic cables are used as transmission lines.

Figure 11-1 illustrates the equivalent circuit of an RF transmission line. It shows that a transmission line exhibits shunt capacitance and series inductance in addition to losses represented by the series resistance and shunt conductance, each

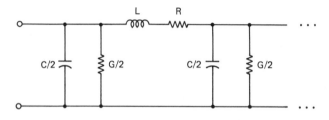

Figure 11-1 Equivalent circuit of RF transmission line.

specified per unit length of line and normally uniformly distributed throughout that length. In many practical transmission lines, the losses are minimal, and the transmission line appears as repetitive sections of shunt capacitance and series inductance.

The most important parameters that characterize a transmission line are its characteristic impedance, Z_o, its attenuation, and its power-handling capability. The characteristic impedance is essentially constant over the useful operating range of the transmission line, while the attenuation and the power-handling capability vary with frequency. The characteristic impedance of a transmission line, which can be modeled by the equivalent circuit in Figure 11-1, is

$$Z_o = \sqrt{\frac{R + j\omega L}{G + j\omega C}}$$

Z_o is in general a complex impedance and is a function of the frequency ω, but for a lossless line it reduces to a real impedance that is independent of frequency.

If a transmission line is terminated with a load different from its characteristic impedance (called a mismatch), a reflection will occur at the load end when energy is sent down the line from a source at the other end. These reflections, which travel back toward the source generator, may adversely affect the characteristics of the source (as well as cause problems on the line), and thus normally must be minimized. The rms voltage along a lossless transmission line with a matched load is constant; however, if a mismatch exists between the transmission line and the load, the voltage will vary in a periodic fashion along the line with a period equal to one-half the wavelength λ. Standing waves are said to exist on a mismatched transmission line; a measure of the mismatch is the voltage standing wave ratio (VSWR), which is defined as

$$\text{VSWR} = \frac{E_{\text{max}}}{E_{\text{min}}} = \frac{E_i + E_r}{E_i - E_r}$$

In this expression, E_{max} is the maximum voltage on the transmission line, E_{min} is the minimum voltage on the line, E_i is the incident voltage magnitude, and E_r is the reflected voltage magnitude. A VSWR of 1.0 indicates a perfect match.

Example 11-1

A transmission line is terminated with a load and driven by a 10 MHz oscillator with an output of 5 volts (rms). The maximum voltage measured along the line is 12 volts (rms), and the minimum voltage measured is 8 volts (rms). Calculate the VSWR and the magnitude of the reflected voltage on this line.

Solution

$$\text{VSWR} = \frac{E_{\text{max}}}{E_{\text{min}}} = \frac{12}{8} = 1.5$$

$$\text{VSWR} = \frac{E_i + E_r}{E_i - E_r}$$

$$1.5 = \frac{5 + E_r}{5 - E_r}$$

$$E_r = 1 \text{ volt (rms)}$$

The voltage standing wave ratio may also be expressed in terms of a complex reflection coefficient Γ, a number whose absolute value is between 0 and 1. This indicates the fraction of the energy reflected back to the source. The expression for the VSWR in terms of the reflection coefficient is

$$\text{VSWR} = \frac{1 + |\Gamma|}{1 - |\Gamma|}$$

Another quantity used to measure mismatch is the return loss, a measure of the power (not voltage) reflected back toward the source (in dB):

$$\text{return loss} = -20 \log |\Gamma|$$

The reflection coefficient Γ for a matched line is 0.0 corresponding to a return loss of $-\infty$ dB and a VSWR of 1.0.

Example 11-2

A transmission line terminated with a load exhibits a VSWR of 2.5. Calculate the reflection coefficient and the return loss.

Solution

$$\text{VSWR} = \frac{1 + |\Gamma|}{1 - |\Gamma|}$$

$$2.5 = \frac{1 + |\Gamma|}{1 - |\Gamma|}$$

$$|\Gamma| = 0.43$$

$$\text{return loss} = -20 \log |\Gamma| = 7.4 \text{ dB}$$

The reflection coefficient may be related to the characteristic impedance of the transmission line, Z_o and the load impedance Z_L as follows:

$$\Gamma = \frac{Z_L - Z_o}{Z_L + Z_o}$$

Example 11-3

A 50-ohm transmission line is terminated with a load impedance of 80 ohms. Calculate the VSWR on this line.

Solution

$$\Gamma = \frac{Z_L - Z_o}{Z_L + Z_o} = \frac{80 - 50}{80 + 50} = \frac{30}{130} = 0.23$$

$$\text{VSWR} = \frac{1 + |\Gamma|}{1 - |\Gamma|} = \frac{1 + |0.23|}{1 - |0.23|} = 1.6$$

Example 11-4

A 50-ohm transmission line is terminated with a load impedance of 20 ohms. Calculate the VSWR on this line.

Solution

$$\Gamma = \frac{Z_L - Z_o}{Z_L + Z_o} = \frac{20 - 50}{20 + 50} = \frac{-30}{70} = -0.43$$

$$\text{VSWR} = \frac{1 + |\Gamma|}{1 - |\Gamma|} = \frac{1 + 0.43}{1 - 0.43} = 2.5$$

Note that the negative reflection coefficient indicates that the incident and reflected voltages are 180° out of phase.

If a transmission line is terminated in a reactive load (that is, an inductor or capacitor) no power will be absorbed by the load and the VSWR will be infinite. A shorted transmission will exhibit a reflection coefficient of -1.0 (VSWR $= \infty$) and an open line will exhibit a reflection coefficient of 1.0 (VSWR $= \infty$). The complex propagation constant γ of a transmission line is defined as follows:

$$\gamma = \alpha + j\beta = \sqrt{(R + j\omega L)(G + j\omega C)}$$

In this expression, α is the real part of the propagation constant and is called the attenuation constant, while β is the imaginary part of the propagation constant and is called the phase constant. R is the series resistance per unit length of the line, L is the series inductance per unit length, G is the shunt conductance per unit length, and C is the shunt capacitance per unit length. The propagation constant may be used to calculate the impedance looking into a length of transmission line terminated in an arbitrary load impedance Z_L:

$$Z_i = Z_o \left[\frac{Z_L + Z_o \tanh \gamma L}{Z_o + Z_L \tanh \gamma L} \right]$$

In this expression, L is the length of the transmission line and tanh is the hyperbolic tangent function, defined as follows:

$$\tanh \gamma = \frac{e^\gamma - e^{-\gamma}}{e^\gamma + e^{-\gamma}}$$

In this expression, γ is a complex number. The exponential function of a complex number is given by Euler's equation:

$$e^{j\beta} = \cos \beta + j \sin \beta$$

If the transmission line is lossless ($G = 0$, $R = 0$), the expression for the input impedance simplifies to

$$Z_i = Z_o \left[\frac{Z_L + jZ_o \tan \beta L}{Z_o + jZ_L \tan \beta L} \right]$$

In the lossless case, the phase constant β is given by:

$$\beta = \omega \sqrt{LC} = \frac{2\pi}{\lambda}$$

In this expression, λ is the wavelength on the line that is shorter than λ_o, the free space wavelength.

The input impedance equations are best solved by computer programs. A number of special cases of the lossless line case should be examined, however. The input impedance of a shorted line is given by

$$Z_i = jZ_o \tan \beta L$$

The input impedance of an open line is given by

$$Z_i = -jZ_o \cot \beta L$$

The input impedance of a line that is an odd number of quarter wavelengths long is

$$Z_i = \frac{Z_o^2}{Z_L}$$

This arrangement, often called a quarter wave transformer, is used in impedance transformation applications.

The input impedance of a line that is an integral number of half wavelengths long is

$$Z_i = Z_L$$

This arrangement provides an input impedance that equals the load impedance for any value of line characteristic impedance.

Example 11-5

A lossless transmission line has a distributed inductance of 10 μH per meter and a distributed shunt capacitance of 500 pF per meter.   Calculate the input impedance of a 4-meter length of this line terminated in a short circuit at 10 MHz.

Solution

$$Z_o = \sqrt{\frac{L}{C}} = \sqrt{\frac{10 \times 10^{-6}}{500 \times 10^{-12}}} = 141 \text{ ohms}$$

$$\beta = \omega \sqrt{LC} = 2\pi f \sqrt{LC} = (6.28)(10 \times 10^6) \sqrt{(10 \times 10^{-6})(500 \times 10^{-12})}$$

$$= 4.44 \text{ radians per meter}$$

$$Z_i = jZ_o \tan \beta L = j(141) \tan [(4.44)(4)]$$

$$Z_i = -j270 \text{ ohms}$$

The input impedance is thus a capacitive reactance of 270 ohms.

The characteristics and uses of specific types of RF transmission lines, such as coaxial line, waveguide, and microstrip, will be discussed later.

11.2 MICROSTRIP TECHNIQUES ━━━━━━━━━━━━━━━━━━━━━━━━━━━

Microstrip is a type of transmission line that has proved to be extremely versatile in RF design.   It is used to realize inductive and capacitive elements for VHF, UHF, and microwave amplifiers as well as in impedance matching applications. Microstrip is also used for traditional transmission line applications and in the design of microstrip antennas.   RF components, such as directional couplers, hybrids and circulators, also often utilize microstrip techniques.

Figure 11-2 illustrates the construction of a microstrip line.   In Figure 11-2, the microstrip line is a metallized strip of thickness t, separated from a ground

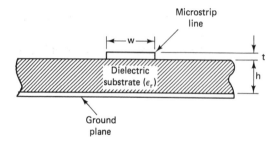

Microstrip
line

Ground
plane

Figure 11-2 Microstrip line construction.

plane by a dielectric substrate material of thickness h, and a relative dielectric constant ε_r. Materials commonly used for the dielectric substrate in microstrip circuits include teflon fiberglass (PTFE) with an $\varepsilon_r = 2.55$ and ceramic loaded PTFE with an $\varepsilon_r = 10.2$. The microstrip line has an air dielectric above it with $\varepsilon_r = 1$, so it is necessary to calculate an effective dielectric constant, ε_r', for microstrip design equations. Two expressions are necessary for ε_r', depending upon the ratio of strip width to dielectric thickness (w/h). For $w/h \leq 1$, the effective dielectric constant is

$$\varepsilon_r' = \frac{\varepsilon_r + 1}{2} + \frac{\varepsilon_r - 1}{2}\left[\left(1 + \frac{12h}{w}\right)^{-\frac{1}{2}} + 0.04\left(1 - \frac{w}{h}\right)^2\right]$$

For $w/h > 1$, the effective dielectric constant is

$$\varepsilon_r' = \frac{\varepsilon_r + 1}{2} + \frac{\varepsilon_r - 1}{2}\left(1 + \frac{12h}{w}\right)^{-\frac{1}{2}}$$

Two expressions are also necessary for the characteristic impedance of the line. For $w/h \leq 1$, the characteristic impedance is

$$Z_o = \frac{60}{\sqrt{\varepsilon_r'}}\ln\left(\frac{8h}{w} + \frac{w}{4h}\right)$$

For $w/h > 1$, the characteristic impedance is

$$Z_o = \frac{\dfrac{120\,\pi}{\sqrt{\varepsilon_r'}}}{\dfrac{w}{h} + 0.667\ln\left(\dfrac{w}{h} + 1.444\right) + 1.393}$$

The above expressions assume negligible strip thickness ($t/h \leq 0.005$).

Normally, in a design situation, the substrate material is known, and it is necessary to compute w/h for a desired characteristic impedance. This may be accomplished by iterative techniques on a computer, or the following closed-form expressions may be used.   For $w/h \leq 2$, the w/h ratio is

$$\frac{w}{h} = \frac{8 \exp\left[\frac{Z_o}{60}\left(\frac{\varepsilon_r + 1}{2}\right)^{\frac{1}{2}} + \frac{\varepsilon_r - 1}{\varepsilon_r + 1}\left(0.23 + \frac{0.11}{\varepsilon_r}\right)\right]}{\exp\left[\frac{Z_o}{30}\left(\frac{\varepsilon_r + 1}{2}\right)^{\frac{1}{2}} + \frac{2(\varepsilon_r - 1)}{(\varepsilon_r + 1)}\left(0.23 + \frac{0.11}{\varepsilon_r}\right)^{-2}\right]}$$

For $w/h > 2$, the w/h ratio is

$$\frac{w}{h} = \frac{2}{\pi}\left[\frac{377\pi}{2Z_o\varepsilon_r^{1/2}} - 1 - \ln\left(\frac{377\pi}{Z_o\varepsilon_r^{1/2}} - 1\right) + \frac{\varepsilon_r - 1}{2\varepsilon_r}\left\{\ln\left(\frac{377\pi}{2Z_o\varepsilon_r^{1/2}} - 1\right) + 0.39 - \frac{0.61}{\varepsilon_r}\right\}\right]$$

Since the above equations separate at $w/h = 2$, and since normally it is not known *a priori* which one to use, solve for the characteristic impedance for $w/h = 2$, and decide which equation to use by determining whether the desired Z_o is higher or lower than the computed Z_o at $w/h = 2$.   Note that Z_o decreases as w/h increases. Also, note that these equations are in terms of ε_r, not ε_r'.

Example 11-6

Design a 50-ohm microstrip transmission line on a 0.064 cm ceramic loaded PTFE substrate.

Solution

$$\varepsilon_r = 10.2 \qquad h = 0.064\,\text{cm}$$

The effective dielectric constant for $w/h = 2$ is

$$\varepsilon_r' = \frac{\varepsilon_r + 1}{2} + \frac{\varepsilon_r - 1}{2}\left(1 + \frac{12h}{w}\right)^{-\frac{1}{2}}$$

$$\varepsilon_r' = \frac{11.2}{2} + \frac{9.2}{2}(1 + 6)^{-\frac{1}{2}} = 7.3$$

The characteristic impedance for $w/h = 2$ is

$$Z_o = \frac{\dfrac{120}{\sqrt{\varepsilon_r'}}}{\dfrac{w}{h} + 0.667\ln\left(\dfrac{w}{h} + 1.444\right) + 1.393}$$

$$Z_o = \frac{\dfrac{(120)(3.14)}{\sqrt{7.3}}}{2 + 0.667\ln(2 + 1.444) + 1.393} = 33.1\,\text{ohms}$$

Since the desired characteristic impedance is larger than the computed value, w/h must be smaller than 2. The following expression is the appropriate one for this case:

$$\frac{w}{h} = \frac{8\exp\left[\frac{Z_o}{60}\left(\frac{\varepsilon_r+1}{2}\right)^{\frac{1}{2}} + \frac{\varepsilon_r-1}{\varepsilon_r+1}\left(0.23 + \frac{0.11}{\varepsilon_r}\right)\right]}{\exp\left[\frac{Z_o}{30}\left(\frac{\varepsilon_r+1}{2}\right)^{\frac{1}{2}} + \frac{2\varepsilon_r-1}{\varepsilon_r+1}\left(0.23 + \frac{0.11}{\varepsilon_r}\right)^{-2}\right]}$$

$$\frac{w}{h} = \frac{8\exp\left[\frac{50}{60}\left(\frac{10.2+1}{2}\right)^{\frac{1}{2}} + \frac{10.2-1}{10.2+1}\left(0.23 + \frac{0.11}{10.2}\right)\right]}{\exp\left[\frac{50}{30}\left(\frac{10.2+1}{2}\right)^{\frac{1}{2}} + \frac{2(10.2-1)}{10.2+1}\left(0.23 + \frac{0.11}{10.2}\right)\right]^{-2}} = 0.94$$

Since $h = 0.064$ cm, w is $0.94\ h = 0.06$ cm.
The very narrow width is the result of using a high dielectric constant substrate.

Microstrip lines are often used to implement inductors and capacitors. Recall that the input impedance of a shorted transmission line is given by

$$Z_i = jZ_o \tan \beta L$$

In this expression, the phase constant β is

$$\beta = \frac{2\pi}{\lambda}$$

In this expression, λ is the wavelength in the microstrip line; it is related to λ_o, the free space wavelength by

$$\lambda = \frac{\lambda_o}{\sqrt{\varepsilon_r'}}$$

The free space wavelength is related to the frequency by

$$\lambda_o = \frac{3 \times 10^8}{f}$$

Examination of the expression for the input impedance of a shorted line indicates that such a line is equivalent to an inductor for small values of βL.

Example 11-7

Design a microstrip inductor with a reactance of 30 ohms at 1 GHz. Use a PTFE substrate with $\varepsilon_r = 2.55$ and a thickness of 0.076 cm.

Solution

The free space wavelength, λ_o, is

$$\lambda_o = \frac{3 \times 10^8}{f} = \frac{3 \times 10^8}{1 \times 10^9} = 3 \times 10^{-1} \text{ m} = 30 \text{ cm}$$

The effective dielectric constant can be calculated by assuming a value for w/h. Assuming $w/h = 2$,

$$\varepsilon_r' = \frac{\varepsilon_r + 1}{2} + \frac{\varepsilon_r - 1}{2} \left(1 + \frac{12h}{w} \right)^{-\frac{1}{2}}$$

$$\varepsilon_r' = \frac{2.55 + 1}{2} + \frac{2.55 - 1}{2} \left(1 + \frac{12}{2} \right)^{-\frac{1}{2}} = 2.1$$

The wavelength on the line is then

$$\lambda = \frac{\lambda_o}{\sqrt{\varepsilon_r'}} = \frac{3 \times 10^{-1}}{\sqrt{2.1}} = 2.1 \times 10^{-1} \text{ m} = 21 \text{ cm}$$

The characteristic impedance for a microstrip line with $w/h = 2$ is

$$Z_o = \frac{\dfrac{120\pi}{\sqrt{\varepsilon_r'}}}{\dfrac{w}{h} + 0.667 \ln \left(\dfrac{w}{h} + 1.444 \right) + 1.393}$$

$$Z_o = \frac{\dfrac{(120)(3.14)}{\sqrt{2.1}}}{2 + 0.667 \ln (2 + 1.444) + 1.393} = 61.8 \text{ ohms}$$

The phase constant β is

$$\beta = \frac{2\pi}{\lambda} = \frac{6.28}{2.1 \times 10^{-1}} = 29.9$$

The input impedance of a shorted microstrip line is

$$Z_i = jZ_o \tan \beta L$$

A Z_i of $j\,30$ results in

$$30 = Z_o \tan \beta L$$

$$30 = (61.8) \tan 29.9\, L$$

$$29.9L = \tan^{-1} \left(\frac{30}{61.8} \right) = \tan^{-1} 0.49 = 0.45$$

$$L = \frac{0.45}{29.9} = 0.015 \text{ m} = 1.5 \text{ cm}$$

Since w/h was set at 2, the width of the microstrip line is

$$w = 2h = 2(0.076 \text{ cm}) = 0.152 \text{ cm}$$

The microstrip inductor is thus realized with a shorted strip of width 0.152 cm and length 1.5 cm.

An open circuit microstrip line may be used to implement a capacitive reactance, since its impedance is

$$Z_i = -jZ_o \cot \beta L$$

Microstrip is also used to transform real values of impedance with quarter wave lines. Recall that the input impedance of a $\lambda/4$ line is

$$Z_i = \frac{Z_o^2}{Z_L} \text{ or } Z_o = \sqrt{Z_i Z_L}$$

Example 11-8

A microwave transistor whose output impedance has been tuned to $25 + j0$ ohms at 2 GHz must be matched to a 50 ohm line. Design an appropriate microstrip quarter wave line, if a PTFE substrate with a thickness of 0.076 cm and $\varepsilon_r = 2.55$ is used.

Solution

The characteristic impedance of the $\lambda/4$ line must be

$$Z_o = \sqrt{Z_i Z_L} = \sqrt{(25)(50)} = 35.4 \text{ ohms}$$

The length of the $\lambda/4$ line is:

$$L = \frac{\lambda}{4} = \frac{\lambda_o}{4\sqrt{\varepsilon_r'}}$$

From Example 11-7, since a $w/h = 2$ for the same substrate resulted in $Z_o = 61.8$ ohms, 35.4 ohms will require $w/h > 2$. Also from this example, $\varepsilon_r' = 2.1$. λ_o is 15 cm.

$$L = \frac{0.15}{4\sqrt{2.1}} = 0.026 \text{ m} = 2.6 \text{ cm}$$

w/h for a characteristic impedance of 35.4 ohms is

$$\frac{w}{h} = \frac{2}{\pi} \left[\frac{377\pi}{2Z_o \varepsilon_r^{1/2}} - 1 - \ln \left(\frac{377\pi}{Z_o \varepsilon_r^{1/2}} - 1 \right) + \right.$$

$$\left. \frac{\varepsilon_r - 1}{2\varepsilon_r} \left\{ \ln \left(\frac{377\pi}{2Z_o \varepsilon_r^{1/2}} - 1 \right) + 0.39 - \frac{0.61}{\varepsilon_r} \right\} \right]$$

$$
= \frac{2}{3.14} \left[\frac{(377)(3.14)}{(2)(35.4)\sqrt{2.1}} - 1 - \ln\left(\frac{(377)(3.14)}{(35.4)\sqrt{2.1}} - 1\right) + \right.
$$

$$
\left. \frac{2.1 - 1}{2(2.1)} \left\{ \ln\left(\frac{(377)(3.14)}{(2)(35.4)\sqrt{2.1}} - 1\right) + 0.39 + \frac{0.61}{2.1} \right\} \right]
$$

$$
= 4.6
$$

The width of the microstrip line is

$$
w = 4.6h = (4.6)(.076) \text{ cm} = 0.35 \text{ cm}
$$

The length of the $\lambda/4$ line is 2.6 cm.

The techniques discussed in this section are applicable primarily to RF amplifier design. Microstrip antennas and other RF components implemented with microstrip will be discussed in later chapters.

11.3 THE SMITH CHART

The Smith chart is a computational aid for evaluating complex transmission line equations. It was first published by Philip H. Smith in the January 1939 issue of *Electronics*. The widespread use of desktop computers has made many applications of the Smith chart obsolete, but the chart is still widely used to display complex impedance or admittance values in semiconductor data books and is sometimes used as a display overlay in impedance measuring instruments. This section has been written to acquaint the reader with the chart, so that it can be used to present data in later sections.

The Smith chart is illustrated in Figure 11-3. Examination of this figure reveals that the Smith chart is based upon two sets of orthogonal circles. The sets of circles are used to represent impedances that are expressed in the rectangular form:

$$
Z = R \pm jX
$$

All of the impedances on a Smith chart are normalized by the characteristic impedance of the transmission line or by a standard real impedance level, such as 50 ohms.

The horizontal line passing through the center of the chart is the pure resistance line. Any point on this line represents an impedance that has only a resistive component (i.e., the reactive component equals zero). The values along this line range from a normalized value of zero resistance on the left to infinity on the right. Note that the center of the line is a normalized resistance of 1.0, which, of course, would be equal to the characteristic impedance Z_o. The circles passing through points on the pure resistance line are known as

IMPEDANCE OR ADMITTANCE COORDINATES

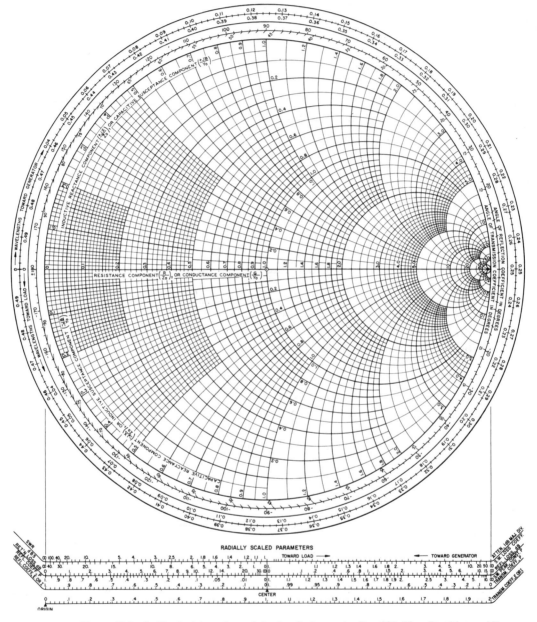

Figure 11-3 Smith chart (courtesy of Analog Instruments, Box 808, New Providence, NJ 07974).

constant resistance circles.   Note that all circles that pass through the pure resistance line to the left of point 2 are not closed at the infinity end of the chart.   This is done to avoid the confusion that would result if all of the grid circles converged to the area at the right of the chart.   Incomplete circles may be estimated or completed with a compass.

The outside circle on the Smith chart is known as the pure reactance circle. A value of reactance may be either inductive or capacitive.   An inductive reactance appears above the pure resistance line and a capacitive reactance appears below the line.   All reactance values are also normalized by the characteristic impedance Z_o.   The circles passing through points on the pure reactance circle are known as constant reactance circles.

Another useful circular scale on the Smith chart is the one labelled "Angle of Reflection Coefficient in Degrees."   This circle lies outside of the pure reactance circle and has a value of 180° at the left side of the chart and a value of 0° at the right.   This circle is used when impedances or other parameters are expressed in polar form (magnitude and angle).

The remaining scales on the Smith chart relate to transmission line calculations which will not be covered here, since they have been essentially replaced by computer methods.

Example 11-9

Plot the following impedances on a Smith chart normalized to $Z_o = 50$ ohms:

$$Z_1 = 200 + j0$$

$$Z_2 = 30 + j60$$

$$Z_3 = 60 - j100$$

$$Z_4 = 0 + j150$$

Solution

The normalized impedances are

$$Z_1 = 4 + j0$$

$$Z_2 = 0.6 + j1.2$$

$$Z_3 = 1.2 - j2$$

$$Z_4 = 0 + j3$$

The impedances are plotted on the Smith chart in Figure 11-4.

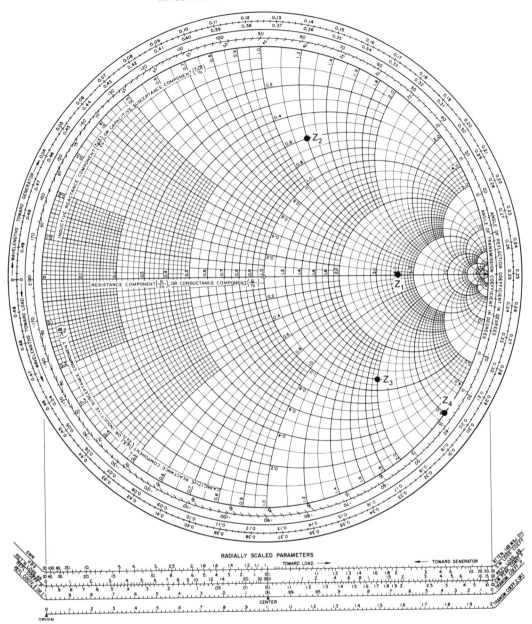

Figure 11-4 Solution to Example 11-6.

11.4 SCATTERING PARAMETERS

Active devices, such as transistors used in RF applications, are normally characterized by a set of two-port parameters. The most common sets of parameters used at frequencies below VHF are the *y*-parameters, *h*-parameters or *z*-param-

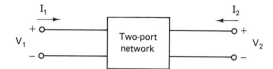

Figure 11-5 Two-port network.

eters.   Figure 11-5 illustrates a two-port network with its input and output voltage and current directions defined.

To illustrate the usefulness of two-port parameters, consider the *h*-parameters that allow the terminal characteristics of a two-port network to be described by the following equations:

$$V_1 = h_{11}I_1 + h_{12}V_2$$

$$I_2 = h_{21}I_1 + h_{22}V_2$$

In the above expressions, h_{11}, h_{12}, h_{21} and h_{22} are known as the *h*-parameters for the network.   Frequently, the *h*-parameters are used to characterize low-frequency transistors.   Parameter sets may also be used to generate equivalent circuits for the network under study.   Figure 11-6 illustrates the equivalent circuit for the *h*-parameters.

In Figure 11-6, $h_{12}V_2$ represents a controlled voltage source and $h_{21}I_1$ represents a controlled current source.   The *h*-parameters of a two-port network are measured as follows:

$$h_{11} = \left. \frac{V_1}{I_1} \right|_{V_2 = 0 \text{ (output shorted)}}$$

$$h_{12} = \left. \frac{V_1}{V_2} \right|_{I_1 = 0 \text{ (input open)}}$$

$$h_{21} = \left. \frac{I_2}{I_1} \right|_{V_2 = 0 \text{ (output shorted)}}$$

$$h_{22} = \left. \frac{I_2}{V_2} \right|_{I_1 = 0 \text{ (input open)}}$$

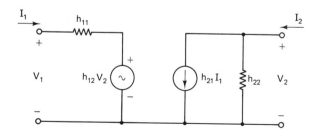

Figure 11-6 *h*-parameter equivalent circuit.

As an example of the above expressions, h_{11} of a transistor or other active device would be measured by shorting the output of the two-port network, applying a known voltage V_1 to the input and measuring the resulting current I_1. The parameter h_{11} would then be the ratio of V_1 to I_1.

The values of h-, y-, and z-parameters are all found by open and short circuit measurements such as those discussed above. Open and short circuit measurements are not practical at VHF frequencies and above, however. Amplitudes vary as a function of transmission line length, short and open circuits are difficult to achieve over a wide frequency range, and active devices are often unstable under open or shorted conditions at these frequencies.

The remedy is to develop a set of parameters based upon incident and reflected wave amplitudes rather than on voltages and currents. Such a set of parameters is called the scattering parameters or more often simply the S-parameters. S-parameters are actually transmission gains and reflection coefficients. Figure 11-7 illustrates, via the use of a signal flowgraph, the relationship between the S-parameters and the incident and reflected wave amplitudes of a two-port network.

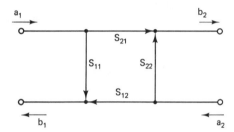

Figure 11-7 S-parameter relationships.

In Figure 11-7, a_1 is the signal into port 1, b_1 the signal out of port 1, a_2 the signal into port 2, and b_2 the signal out of port 2. In Figure 11-7, b_1 may be thought of as the reflected signal at the input and a_2 as the reflected signal from the output load. The terminal characteristics of the two-port network are described by the following equations:

$$b_1 = S_{11}a_1 + S_{12}a_2$$

$$b_2 = S_{21}a_1 + S_{22}a_2$$

TABLE 11-1   HXTR-6104 S-PARAMETERS

f(GHz)	S_{11}	S_{21}	S_{12}	S_{22}
2.0	0.50 $\angle -151°$	2.93 $\angle 69°$	0.05 $\angle 31°$	0.72 $\angle -43°$
2.5	0.50 $\angle -169°$	2.45 $\angle 55°$	0.05 $\angle 31°$	0.69 $\angle -51°$
3.0	0.49 $\angle 175°$	2.12 $\angle 42°$	0.06 $\angle 33°$	0.68 $\angle -57°$
3.5	0.54 $\angle 165°$	1.87 $\angle 29°$	0.06 $\angle 35°$	0.65 $\angle -68°$
4.0	0.52 $\angle 156°$	1.67 $\angle 19°$	0.06 $\angle 37°$	0.68 $\angle -76°$

IMPEDANCE OR ADMITTANCE COORDINATES

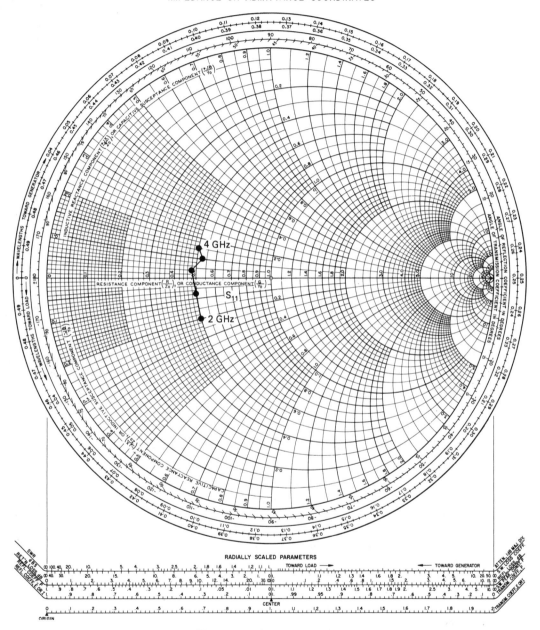

Figure 11-8 S_{11} for HXTR-6104 transistor.

The *S*-parameters of a two-port network are measured as follows:

$$S_{11} = \left. \frac{b_1}{a_1} \right|_{a_2 = 0 \text{ (output terminated in } Z_o)}$$

$$S_{21} = \left. \frac{b_2}{a_1} \right|_{a_2 = 0 \text{ (output terminated in } Z_o)}$$

$$S_{12} = \left. \frac{b_1}{a_2} \right|_{a_1 = 0 \text{ (input terminated in } Z_o)}$$

$$S_{22} = \left. \frac{b_2}{a_2} \right|_{a_1 = 0 \text{ (input terminated in } Z_o)}$$

The measurement of a device's or a network's *S*-parameters requires terminations of value Z_o (normally 50 ohms), which are easily achieved at VHF and higher frequencies.

The parameter S_{11} is called the input reflection coefficient, S_{21} is the forward transmission gain or loss, S_{12} is the reverse transmission coefficient, and S_{22} is the output reflection coefficient.

The *S*-parameters are widely used in the design of low noise RF amplifiers. Table 11-1 on page 202 tabulates the *S*-parameters of the Hewlett-Packard HXTR-6104 bipolar transistor from 2 to 4 GHz.

The *S*-parameters are also sometimes presented on Smith charts in device data sheets. Figure 11-8 on page 203 illustrates the variation of S_{11} from 2 GHz to 4 GHz for the HXTR-6104 transistor on a Smith chart. Data sheets may also provide *S*-parameter data in rectangular form.

EXERCISES ━━━━━━━━━━━━━━━━━━━━━━━━━━━━━━━━━━━━

1. The maximum voltage measured along a transmission line is 9 volts and the minimum voltage is 4 volts. Calculate the VSWR on the line.
2. Calculate the reflection coefficient and return loss for a transmission line with a VSWR of 1.8.
3. A 75-ohm transmission is terminated with a load of 50 ohms. Calculate the reflection coefficient, VSWR, and return loss for this line.
4. A lossless transmission line has a distributed inductance of 8 μH per meter and a distributed shunt capacitance of 300 pF per meter. Calculate the input impedance of a 1-meter section of this line terminated in an open circuit at 70 MHz.
5. Design a 50-ohm microstrip transmission line on a 0.064 cm substrate with a dielectric constant of 2.33.
6. Design a 1 pF microstrip capacitor using microstrip techniques at 1 GHz. Assume a 0.076 cm substrate with $\varepsilon_r = 2.55$.

7. Design a quarter wave microstrip matching transformer to match 35 ohms to 50 ohms at 3 GHz.   Assume a 0.076 cm substrate with a dielectric constant of 2.33.

8. Plot the following impedances (normalized to 50 ohms) on a Smith chart:

$$Z_1 = 650 - j860$$

$$Z_2 = 0 + j100$$

$$Z_3 = 100 + j0$$

$$Z_4 = 60 + j60$$

9. Plot the *S*-parameters for the HXTR-6104 transistor from 2 to 4 GHz on a rectangular grid.

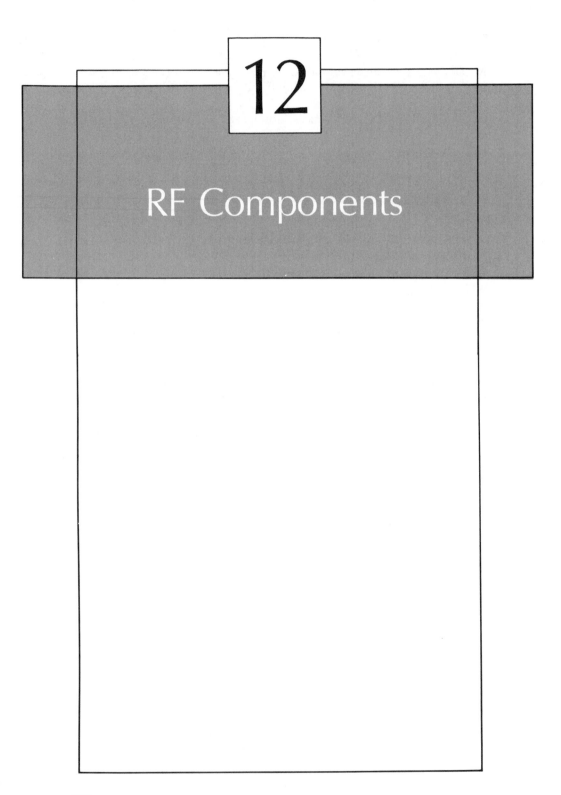

12

RF Components

Most current communications systems utilize a number of prepackaged components in their RF sections, including directional couplers, quadrature hybrids, circulators, isolators, attenuators, and microstrip filters.   These components generally are designed using microstrip techniques (covered in Chapter 11) or with stripline.   A stripline circuit is similar to a microstrip circuit in that it is constructed with etched lines on a dielectric substrate backed by a ground plane, but differs in that it also has a dielectric substrate with ground plane over the etched lines in a "sandwich" arrangement.   Stripline circuits provide shielding superior to that offered by microstrip circuits.   Each component to be discussed in this chapter performs a specific function in the RF section of a communications transmitter or receiver.

12.1 DIRECTIONAL COUPLERS

A directional coupler provides a sample of the energy flowing along an RF transmission line in either the forward or reverse direction.   Directional couplers are implemented with a primary section of microstrip or stripline and a secondary strip either side-coupled or broadside-coupled to the primary section.   RF energy is coupled from the primary strip to the secondary strip with no physical connection between the two lines.   The amount of energy coupled into the secondary line depends upon the spacing between the lines.   The directional property of a coupler is derived from the length of the secondary line, which is designed to be a quarter wavelength long.   A quarter wavelength line exhibits a low impedance at one end and a high impedance at the other.   RF energy is readily coupled into the secondary line at the coupling end, while very little RF energy is coupled at the opposite end. Figure 12-1 illustrates a directional coupler.

In Figure 12-1, if RF energy is applied to port 1 of the directional coupler, the majority of the energy will travel to port 3 with a small amount of insertion loss, and a sample of the energy will appear at port 2.   Port 4 will receive minimal energy.   If the RF energy is applied to port 3, port 1 will receive most of the energy with a fraction coupled to port 4 and minimal energy at port 2.   Port 1 is normally called the input port, port 2 the coupled port, port 3 the straight-through port and port 4 the isolated port.   The isolated port in a directional coupler is often internally terminated with a resistor (normally 50 ohms); the component then is known as a three-port directional coupler.

A number of specifications characterize a directional coupler.   The frequency range of a coupler specifies the frequency band for which it was designed and for

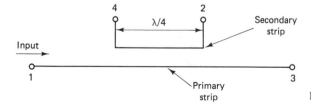

Figure 12-1   Directional coupler.

Figure 12-2 Dual directional coupler.

which the other specifications are valid. Coupling is the amount, expressed in dB, that the power coming out of the coupled port is less than the power into the input port. Common values for coupling are 10 dB, 20 dB or 30 dB, although other values are available. Directivity is a measure of isolation. If power is applied to the input port and the coupled port signal is 10 dB down and the isolated port signal is 45 dB down, the directivity would be 35 dB. Insertion loss is the loss from the input port to the straight-through port due to coupling and line loss. Insertion loss is greatest when coupling is the tightest; normally it is less than 1 dB.

Directional couplers are also specified by their power-handling capability, their voltage standing wave ratio (VSWR) and their coupling deviation.

A dual directional coupler has two couplers integrated back-to-back in a single package. Figure 12-2 illustrates a dual directional coupler. A dual directional coupler has its two isolated ports terminated internally.

Directional couplers are often inserted in systems when it is necessary to measure the power level or frequency of an RF signal. The insertion loss of the coupler is normally negligible and thus has little effect upon the system operation. Directional couplers are also used when another part of the system requires a sample of an RF signal; a dual directional coupler is often used to measure forward and reverse power simultaneously in order to compute the VSWR.

12.2 QUADRATURE HYBRIDS

A quadrature hybrid is a directional coupler whose outputs are equal in magnitude and differ in phase by 90°. Figure 12-3 illustrates a quadrature hybrid.

In Figure 12-3, the two output ports are isolated from each other. The input port is also isolated from the isolated port, which is either internally or externally terminated in the characteristic impedance of the device (normally 50 ohms). The

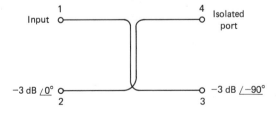

Figure 12-3 Quadrature hybrid.

power available at port 2 is in phase with the input and down by 3 dB. The power at port 3 lags the input by 90° and is also down by 3 dB.

Quadrature hybrids are characterized by parameters such as operating frequency range, isolation between ports, insertion loss, VSWR, and amplitude and phase balance between output ports.

Quadrature hybrids may be used in modulation and demodulation schemes where in-phase and quadrature carrier signals are required. They are also used in mixer circuits and in RF amplifiers when the power from two active devices must be combined.

12.3 CIRCULATORS

A circulator is a three-port RF component that passes RF energy from port to port in one direction. In the circulator illustrated in Figure 12-4, power applied to port 1 will appear at port 2 minus a small insertion loss. Reflected power from port 2 will travel to port 3, not back to port 1. Either port 2 or port 3 could be used as an input with the same result—the power would only travel in a forward direction.

A circulator is constructed from a dielectric substrate with three striplines physically shorted together and surrounded top and bottom by ferrite rings and magnets. Its action depends upon the interaction of the RF fields and the magnetic fields in the device; it differs from a directional coupler in that, while both devices exhibit RF isolation, the circulator has its three ports shorted and thus exhibits no isolation between its ports at DC. Proper operation of a circulator is highly dependent upon the presence of a matched load at each of its ports.

Circulators are characterized by frequency range, insertion loss, isolation between ports, VSWR and power capability.

The most common application of a circulator is as a diplexer, which allows a single antenna to simultaneously feed a receiver and be fed from a transmitter. In Figure 12-5, the transmitter power applied to port 1 will exit at port 2 to the antenna. Received power from the antenna will enter port 2 and exit at port 3 to the receiver. In this application, proper matching is especially important, because isolation must be maximized.

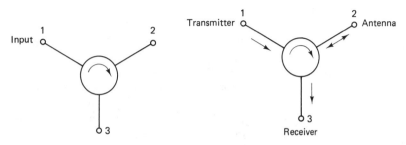

Figure 12-4 Circulator.   **Figure 12-5** Circulator used as a diplexer.

12.4 ISOLATORS

An isolator is essentially a circulator with the third port of the device internally terminated with its characteristic impedance (normally 50 ohms). Thus, it is a two-port device that allows power to flow only in one direction. The purpose of an isolator is to provide isolation between RF circuits, effectively eliminating the undesirable effects of a high VSWR, which might occur as a result of impedance mismatches. As an example, a high VSWR of 5:1 between a source and a load can be reduced to approximately 1.15:1 by inserting a 20 dB isolator between the source and the load.

Isolators are characterized by the same parameters as circulators, differing only by the internal termination for the isolator.

Isolators are used whenever a mismatch between two RF circuits can cause undesirable effects. Isolators are also useful in reducing the amount of local oscillator radiation back through the antenna in a receiving system or the effects of a changing load on the frequency of an oscillator in a transmitter.

12.5 ATTENUATORS

Attenuators reduce the power level of an RF signal. They are used in communications systems to adjust the signal levels for maximum efficiency of operation of subsequent circuits and are also widely used in the testing of communications circuits to reduce power levels to a value appropriate to the circuit being tested or the measuring instrument in use.

RF attenuators are realized with either a "T" or a "π" configuration of resistive elements. At higher RF frequencies, the resistive elements are usually fabricated by depositing a nichrome alloy film on a ceramic substrate.

Attenuators are available in either fixed or variable configurations. Variable attenuators are normally used in the design and development phase or in testing, while fixed attenuators are usually used in production models of the system. Attenuators are characterized by frequency range, attenuation, accuracy, VSWR, and maximum input power.

12.6 MICROSTRIP FILTERS

RF filters are available to the communications system designer as prepackaged components. They are normally implemented with microstrip or stripline techniques in conjunction with passive filter design methods.

Microstrip techniques may be used to implement low pass, high pass, band pass, and band reject filters with Butterworth, Chebyshev, elliptic, or any of the other response characteristics.

Band pass filters available to the RF circuit designer are normally specified by passband, insertion loss, ripple, shape factor, and rejection.   Low pass and high pass filters normally are specified for a cutoff frequency rather than a passband, and band reject filters are specified for a rejection band.

It is important to note that the passband of a filter is not equivalent to its bandwidth.   Bandwidth normally specifies the frequency range over which response is no more than a specified value (for example, 3 dB down), while passband is the frequency range over which the signal is attenuated by no more than a specified insertion loss.   The response curve of a filter is often described by a quantity known as shape factor, which is the ratio of the 60 dB bandwidth to the 3 dB bandwidth in a filter.   The shape factor is also sometimes specified at bandwidths other than 60 dB (such as 30 dB).   The rejection of a filter is simply the amount of attenuation in dB at a specified frequency on the "skirts" of the response curve.

Example 12-1

A microstrip filter with a passband of 2 GHz to 4 GHz has a 3 dB bandwidth of 3 GHz.   The bandwidth of the filter at 60 dB down is 7.5 GHz.   Calculate the shape factor of this filter.

Solution

$$BW_{3\ dB} = 3\ \text{GHz} \qquad BW_{60\ dB} = 7.5\ \text{GHz}$$

$$\text{Shape factor} = \frac{BW_{60\ dB}}{BW_{3\ dB}} = \frac{7.5}{3} = 2.5$$

EXERCISES

1. A 10 dBm, 18 GHz signal is applied to a 10 dB directional coupler with a directivity of 40 dB and an insertion loss of 0.5 dB.   Calculate the power at the coupled port and the isolated port in milliwatts.

2. A 20 dB dual directional coupler inserted between an amplifier and a mixer in a 12 GHz receiving system indicates a forward power flow of 0 dBm and a reverse power flow of −10 dBm.   Calculate the VSWR of the line and the reflection coefficient of the mixer.

3. A 3 dB quadrature hybrid is driven by a signal at 4 GHz of 10 dBm.   Calculate the power at the two outputs in milliwatts and indicate all of the phase relationships.

4. A 1 milliwatt signal is applied to port 1 of a circulator that has a specified isolation of 20 dB.   Calculate the power out of the other two ports in dBm.

5. Calculate the shape factor of a microstrip filter with a passband of 4 GHz to 8 GHz and a 3 dB bandwidth of 5 GHz.   The bandwidth of the filter at the 60 dB down point is 8 GHz.

13

RF Frequency
Generation Circuits

RF frequency generation circuits are common to both communications receivers and communications transmitters. Nearly all receivers in current use utilize the superheterodyne principle, whereby the received carrier frequency is converted to a lower intermediate frequency, at which the majority of amplification and demodulation takes place. A superheterodyne receiver requires an RF source, the local oscillator, which is mixed with the carrier to produce the intermediate frequency (IF) signal. In some receivers, the down conversion process is accomplished in several steps, resulting in multiple intermediate frequencies and thus requiring additional oscillators. All transmitters require an initial source of RF energy, which is amplified and modulated to produce the final output. In some transmitters, the superheterodyne principle is employed, and several oscillators may be required.

Another important component in both receivers and transmitters is the phase locked loop. A key element in the PLL is the voltage controlled oscillator. The phase locked loop is the heart of the digital frequency synthesizer, itself an extremely important RF frequency generation circuit.

Although mixers and frequency multipliers generate no RF energy of their own, they play an essential role in superheterodyne frequency conversion and/or translation, and thus are important elements of most receivers and transmitters.

13.1 OSCILLATOR CIRCUITS

An oscillator is essentially an amplifier with a frequency-selective positive feedback network. This circuit configuration will produce RF energy at a frequency determined by the feedback network, and provided certain circuit constraints are met the output waveform will be sinusoidal. Figure 13-1 illustrates a simplified block diagram of an oscillator. In this figure, the gain of the amplifier as a function of frequency is $A(f)$, and the gain of the feedback network as a function of frequency is $\beta(f)$. The equations describing the system illustrated in Figure 13-1 are

$$v_e(t) = v_i(t) + \beta(f)v_o(t)$$

$$v_o(t) = A(f)v_e(t)$$

If $v_e(t)$ is eliminated from this set of equations, the following equation is obtained:

$$\frac{v_o(t)}{A(f)} = v_i(t) + \beta(f)v_o(t)$$

This equation may be solved for the system transfer function $v_o(t)/v_i(t)$:

$$\frac{v_o(t)}{v_i(t)} = \frac{A(f)}{1 - \beta(f)A(f)}$$

The system will oscillate at a frequency where the denominator of the transfer

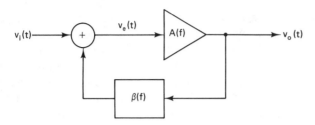

Figure 13-1 Simplified block diagram of an oscillator.

function becomes zero or when

$$\beta(f)A(f) = 1$$

In practice, the frequency behavior of the amplifier in the region where oscillation is desired is flat, so only the frequency response characteristics of the feedback network need to be considered. The feedback network will normally meet the condition for oscillation at only one frequency, which is determined by the value of its components.

Figure 13-2 illustrates the circuit diagram of a common oscillator circuit (a Colpitts oscillator) implemented with a bipolar transistor. The Colpitts oscillator oscillates at a frequency

$$f_o = \frac{1}{2\pi\sqrt{LC}}$$

In this expression, L is the value of the inductance in parallel with the load resistor, R_L, and C is the equivalent capacitance of C_1 and C_2 in series. The oscillations will be sinusoidal if the following two conditions are met:

$$Q = 2\pi f_o C R_L \gg 1 \qquad n = \frac{C_1}{C_1 + C_2} \ll 1$$

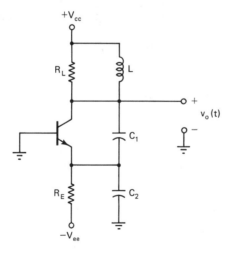

Figure 13-2 Colpitts oscillator.

In these expressions, Q is the quality factor of the circuit (a measure of selectivity), and n is the feedback ratio.

Example 13-1

The Colpitts oscillator of Figure 13-2 has the following component values:

$$L = 5 \text{ } \mu\text{H} \qquad C_2 = 5000 \text{ pF}$$

$$R_L = 10 \text{ K}\Omega \qquad R_E = 20 \text{ K}\Omega$$

$$C_1 = 50 \text{ pF} \qquad V_{cc} = +10 \text{ v}$$

$$V_{ee} = -10 \text{ v}$$

Calculate the frequency at which this circuit will oscillate and determine if the oscillations are sinusoidal.

Solution

$$C = \frac{C_1 C_2}{C_1 + C_2} = \frac{(50)(5000)}{(50 + 5000)} = 49.5 \text{ pF}$$

$$f_o = \frac{1}{2\pi \sqrt{LC}} = \frac{1}{2\pi\sqrt{(5 \times 10^{-6})(49.5 \times 10^{-12})}} = 10.1 \times 10^6 = 10.1 \text{ MHz}$$

$$Q = 2\pi f_o C R_L = (6.28)(10.1 \times 10^6)(49.5 \times 10^{-12})(10 \times 10^3) = 31.4$$

$$n = \frac{C_1}{C_1 + C_2} = \frac{50}{5000 + 50} = 0.01$$

The given circuit will provide sinusoidal oscillations at a frequency of 10.1 MHz.

The Colpitts oscillator may be used to realize a voltage controlled oscillator (VCO) by placing a voltage variable capacitor or varactor across the two series capacitors in Figure 13-2.

The Colpitts oscillator is used only where frequency stability is not very important, because it is affected by changes in transistor output capacitance caused by small changes in supply voltage.

When frequency stability is important in an oscillator, a quartz crystal is used as the frequency-determining element in the circuit. A quartz crystal may exhibit a Q that is up to 1000 times greater than that of an LC circuit. Quartz crystals are available in various "cuts" that have different frequency vs. temperature properties, different frequency ranges of operation, and different relationships between their electrical and mechanical properties. Figure 13-3 illustrates the equivalent circuit of a quartz crystal.

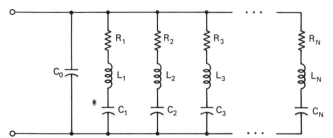

Figure 13-3 Equivalent circuit for quartz crystal.

In Figure 13-3, C_o is the capacitance between the case electrodes. The series RLC circuits correspond to the crystal's fundamental mode of oscillation as well as its odd harmonic modes (which are called overtones). The cut and the mounting of a crystal determine whether it will operate in its fundamental mode or on one of its overtones. In any event, the effective equivalent circuit for one mode consists of the capacitance C_o and one of the series RLC branches at a given frequency of operation.

In a Colpitts oscillator circuit, the crystal is normally inserted in the feedback path, as illustrated in Figure 13-4, where it is utilized in a series-resonant mode of operation (crystals display two resonant frequencies in each mode: series-resonant and parallel-resonant). The L-C resonant circuit in a Colpitts crystal oscillator should be designed for a center frequency equal to the crystal's resonant frequency.

Quartz crystals may be "pulled" in frequency a small amount by placing a capacitor in parallel with the crystal. If the capacitor is a varactor, the frequency of the oscillator may therefore be varied electrically over a small range to produce FM modulation.

A number of other oscillator circuit configurations exist, including the Hartley, tuned-collector, and Miller oscillators.

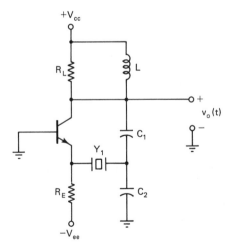

Figure 13-4 Colpitts crystal oscillator.

   Oscillator circuits are available in integrated circuit form up to approximately 250 MHz.   An example is the Plessey SP1648, which operates from either a +5-volt or −5.2-volt supply and is available in a 14-pin dual in-line package.   The SP1648 requires an external inductor and capacitor and operates at frequencies up to 225 MHz.

13.2 DIGITAL FREQUENCY SYNTHESIZERS

Most communications transmitters and receivers must operate on a number of discrete frequencies.   One solution to this requirement is to make one of the capacitors in the frequency-determining feedback network a variable capacitor. The disadvantage of this approach is the lack of precision in setting the frequency and the reduced stability resulting from a variable capacitor.   Another solution is to use a bank of crystals, which can be switched into the oscillator circuit.   The disadvantage here is the cost of the crystals and the physical space they consume. The modern solution to the indicated problems is the digital frequency synthesizer, which is used in virtually all current communications systems where channelized operation is required.   Figure 13-5 illustrates a basic digital frequency synthesizer block diagram.

   The digital frequency synthesizer illustrated in Figure 13-5 is a phase locked loop with a counter (frequency divider) inserted into the feedback loop.   The input to the synthesizer is a stable reference frequency, usually crystal-controlled.   The PLL will lock onto the reference frequency if the output of the VCO is N times the reference frequency, where N is the division factor of the counter.   In most digital frequency synthesizers, the $\div N$ counter is programmable with a digital input word, so N may be changed, resulting in a change in the output frequency.   The spacing between channels in a digital frequency synthesizer equals the reference frequency f_r, since N must change by integer values.

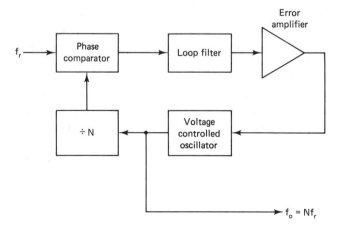

Figure 13-5   Digital frequency synthesizer block diagram.

Example 13-2

A digital frequency synthesizer is designed with a reference frequency of 100 KHz and a programmable counter, which ranges from 16 to 64. Calculate the range of output frequencies, number of channels and channel spacing for this synthesizer.

Solution

$$f_o = Nf_r$$

$$f_{o_1} = (16)(100 \times 10^3) = 1.6 \times 10^6 = 1.6 \text{ MHz}$$

$$f_{o_2} = (64)(100 \times 10^3) = 6.4 \times 10^6 = 6.4 \text{ MHz}$$

$$\text{Number of channels} = 64 - 16 + 1 = 49$$

$$\text{Channel spacing} = f_r = 100 \text{ KHz}$$

The digital frequency synthesizer is capable of providing 49 contiguous channels spaced 100 KHz apart over the 1.6 MHz to 6.4 MHz range.

Digital frequency synthesizers may be made to operate at frequencies higher than the maximum frequency of programmable counters by inserting a second counter before the $\div N$ counter. Such a counter is known as a prescaler and usually has only one or two divide ratios. If a prescaler with a divide ratio of M is inserted into the reference frequency line, fractional frequency spacing may be obtained, since the output frequency in such an arrangement is $f_o = N/M \, f_r$.

In some special applications, more complex synthesizer circuits are used with dual modulus programmable counters and multiple loops.

Digital frequency synthesizers are available on a single chip with both parallel and serial frequency selection capability. An example of an integrated circuit digital frequency synthesizer is the Motorola MC145151-1, which is a CMOS chip with parallel input lines. The MC145151-1 operates from a 3-volt to a 9-volt supply and has a maximum operating frequency of approximately 26 MHz. The chip has a $\div N$ range of 3 to 16383. The MC145151-1 has an on-chip reference oscillator, but requires an external loop filter and VCO.

Voltage controlled oscillators are also available in integrated circuit form. The National Semiconductor LM566 is a VCO that operates from a supply voltage of 10 volts to 24 volts; it is available in an 8-pin dual in-line package. Its maximum operating frequency is 1 MHz.

The switching speed of a digital frequency synthesizer is determined by the transient response of the phase locked loop. PLL transient response was covered in Chapter 9.

13.3 MIXER CIRCUITS

Mixers are widely used in communications systems. The mixer is used in some transmitter circuits, but its primary application is in superheterodyne receivers. A mixer is a device with two input ports and one output port. The frequency of the signal at the output port can be either the sum or the difference of the frequencies of the two signals applied to the two input ports. Mixers are most often used as down converters in receivers where the received modulated carrier signal is mixed with a local oscillator signal to translate it to a lower intermediate frequency (IF) for further amplification, filtering, and demodulation. The IF center frequency is the difference between the carrier frequency of the received signal and the local oscillator frequency. Figure 13-6 illustrates a mixer schematic.

A mixer circuit is basically a multiplier circuit, so the previous discussion on multiplier circuits applies to the mixer. The primary difference between a mixer and a balanced modulator is that the balanced modulator is followed by a low pass filter at baseband, while the mixer is followed by a bandpass filter at the IF.

Virtually all of the mixer circuits used in current communications systems are double-balanced mixers because of their superior performance. Double-balanced mixers may be passive or active. Passive double-balanced mixers utilize a diode ring and transformers, and are available as a component. Active mixers are available in integrated circuit form, and they use internal differential amplifiers to achieve the double-balanced mixer function.

Mixers are characterized by their conversion loss, their isolation between ports, their noise figure, the local oscillator power required, and VSWR at the RF and LO ports.

Conversion loss is the level of the IF output signal as compared to the level of RF input signal, expressed in dB. Conversion loss is a measure of the mixer efficiency. Isolation is a measure of how far down (in dB) an undesired signal is at a given port. Noise figure is an important mixer parameter, since it is a measure of the degradation in the signal-to-noise ratio for a signal passing through the device. The noise figure is expressed in dB. The local oscillator power required for proper mixer operation is a key parameter, since the local oscillator must be designed to supply this level of power. As discussed previously, VSWR is a measure of mismatch between two RF circuits and ideally is 1:1. Mixers with poor VSWRs often require isolators to improve their performance.

An example of a passive double-balanced mixer is the Mini-Circuits Labo-

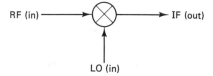

Figure 13-6 Mixer schematic.

ratory (MCL) model MA-1. The MA-1 can operate with LO and RF inputs from 1 to 2500 MHz and an IF from 1 to 1000 MHz. The MA-1 exhibits a conversion loss of 8 dB and has 40 dB of isolation between the LO/RF ports and the LO/IF ports. The local oscillator power required is +10 dBm.

Balanced modulator integrated circuits, such as the Motorola MC1496, may be used as double-balanced mixers, or devices specifically designed to be used as mixers, such as the Plessey SL6440A, may be used. The SL6440A will operate with RF and LO inputs up to 150 MHz, has a conversion loss of 1 dB, and has 40 dB of isolation from the LO to RF ports. The device is available in a 16-pin dual in-line package.

The Signetics NE602 is an integrated circuit that combines a local oscillator and double-balanced mixer on a single chip. The NE602 is available in an 8-pin dual in-line package and operates from a 6-volt supply. The RF input signal frequency is 200 MHz maximum. The internal local oscillator may also be operated up to 200 MHz. The device exhibits a noise figure of 5 dB at 45 MHz.

13.4 FREQUENCY MULTIPLICATION CIRCUITS

A frequency multiplier is a circuit that follows an oscillator stage, or the output of a digital frequency synthesizer, and multiplies its input frequency by an integer, such as 2 (doubler) or 3 (tripler). Frequency multiplication may be implemented by passing the input signal through a non-linear device, such as a diode or transistor amplifier operating non-linearly (class C) to produce harmonics. The output of the multiplier is bandpass filtered at the desired harmonic. A frequency multiplier may also be constructed from an analog multiplier with the input signal applied to both input ports on the device and the output passed through a band pass filter. The efficiency of multiplier circuits is generally low, and it decreases as the order of the multiplication is increased.

Frequency multiplier circuits are often used in FM transmitters, where the FM oscillator is adjusted for a small frequency deviation, which is subsequently multiplied by passing the oscillator output through one or more frequency multiplier circuits. The overall modulation linearity of FM modulation is improved with such a circuit arrangement.

Frequency multiplication circuits are used in the frequency generation portion of VHF, UHF, and microwave transmitters and also sometimes in the local oscillator chain in receivers operating at these frequencies.

EXERCISES

1. A Colpitts oscillator has $L = 500$ nH, $R_L = 10$ KΩ, $C_1 = 10$ pF and $C_2 = 300$ pF. Calculate its oscillation frequency.

2. Design a Colpitts oscillator with practical component values for a frequency of 50 MHz.

3. For a quartz crystal, $C_o = 0.1$ pF, $R_1 = 0.1$ Ω, $L_1 = 50$ nH and $C_1 = 50$ pF. The crystal is cut so that it will oscillate in its fundamental mode. Calculate its frequency of oscillation.

4. A digital frequency synthesizer has a reference frequency of 10 KHz and a fixed $\div 8096$ counter. Calculate the output frequency of this circuit.

5. A digital frequency synthesizer has a reference frequency of 20 KHz and a programmable $\div N$ counter that ranges from 16384 to 65536. Calculate the range of output frequencies, the number of channels, and the channel frequency spacing for this synthesizer.

6. A double-balanced mixer with a conversion loss of 6 dB is used in a receiver with a local oscillator power of 7 dBm. If a -10 dBm signal is applied to the RF input, calculate the signal power at the IF output.

7. A signal with a signal-to-noise ratio of 20 dB is applied to a double-balanced mixer with a conversion loss of 9 dB, a local oscillator power of 7 dBm, and a noise figure of 10 dB. Calculate the signal-to-noise ratio at the IF output of this mixer.

8. A 48 MHz FM signal with a deviation of 5 KHz is passed through a tripler frequency multiplication circuit. Calculate the carrier frequency and the deviation of the FM signal at the output of the multiplier.

14

RF Amplifier Circuits

The RF amplifier is an essential component in communications transmitters and receivers.   In a transmitter, several stages of RF power amplification are required after the frequency generating circuit to bring the output power up to a level suitable for transmission.   The signal level at the input of a receiving system may only be a fraction of a microwatt, which must be amplified to several watts in many cases. A receiver normally utilizes a low-noise amplifier at its input, several stages of high-gain amplification at one or more intermediate frequencies, and a baseband amplifier following the detector circuit.

RF amplifiers are designed to be narrowband (< 10% bandwidth) or wideband, and may be optimized for maximum gain, maximum power, or minimum noise figure.   Small-signal RF amplifier design is usually based upon S-parameter models and microstrip matching networks at the frequencies used in most current communications equipment.

Although specialized design techniques are required in some situations, the RF amplifier can be considered a component, and prepackaged amplifiers are now available for most applications.   Some understanding of design techniques, however, is helpful in the interpretation of data sheets and the proper application of the amplifier modules.

14.1 LOW NOISE AMPLIFIERS

Low noise RF amplifiers are biased for class A operation and usually utilize silicon bipolar or FET transistors up to approximately 2 GHz and gallium arsenide field effect transistors (GaAs FETs) above this frequency.   The type of GaAs FET most often used in low noise RF amplifier applications is the MESFET, a field effect transistor with a metal-semiconductor junction at the gate of the device, called a Schottky barrier.   Gallium arsenide is used as a semiconductor material at the higher frequencies, because it exhibits a higher electron bulk mobility and drift velocity than silicon.   These properties result in lower noise figure, higher gain, and higher cutoff frequencies.   Silicon bipolar transistors and gallium arsenide field effect transistors are best characterized for low-noise applications by the use of S-parameters. Figure 14-1 illustrates a general block diagram of a single-stage RF amplifier.

A common measure of gain in RF amplifier design is the transducer power gain, G_T:

$$G_T = \frac{\text{Power absorbed by load}}{\text{Power available from generator}}$$

The transducer power gain of an RF amplifier in terms of the S-parameters of the device and the source and load reflection coefficients is

$$G_T = \frac{|S_{21}|^2 \, (1 - |\Gamma_s|^2)(1 - |\Gamma_L|^2)}{|(1 - S_{11}\Gamma_s)(1 - S_{22}\,\Gamma_L) - S_{21}S_{12}\,\Gamma_L\,\Gamma_s|^2}$$

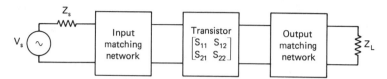

Figure 14-1 Block diagram of single-stage RF amplifier.

In this expression, S_{11}, S_{12}, S_{21} and S_{22} are the scattering parameters of the device, Γ_s is the reflection coefficient of the source, and Γ_L is the load reflection coefficient. The source reflection coefficient is

$$\Gamma_s = \frac{Z_s - Z_o}{Z_s + Z_o}$$

where the characteristic impedance Z_o is normally 50 ohms. The load reflection coefficient is given by

$$\Gamma_L = \frac{Z_L - Z_o}{Z_L + Z_o}$$

The maximum transducer gain will occur when Γ_s equals the complex conjugate of S_{11}, and Γ_L equals the complex conjugate of S_{22}. (The reverse transmission coefficient S_{12} must be negligible for the maximum to occur exactly at the conjugate match points.) The maximum transducer gain is

$$G_{T_{max}} = \frac{|S_{21}|^2}{(1 - |S_{11}|^2)(1 - |S_{22}|^2)}$$

Minimum noise figure in a transistor amplifier will occur when the input is slightly mismatched. The maximum gain for the minimum noise figure condition will occur when the output is conjugate matched. The reflection coefficient at the output for a given source reflection coefficient is

$$\Gamma_o = S_{22} + \frac{S_{12}S_{21}\,\Gamma_s}{1 - \Gamma_s S_{11}}$$

Transistor data sheets (for low-noise devices) give an optimum source reflection coefficient, $\Gamma_{S_{opt}}$, which results in minimum noise figure. The function of the input matching network is to transform the source impedance to this optimum input impedance.

An important consideration in RF amplifier design is stability. An unstable amplifier will oscillate at undesired frequencies and become unusable. An amplifier is said to be unconditionally stable if it will not oscillate for any combination of input and output impedances. For unconditional stability an amplifier must have a stability factor K greater than 1, and $|S_{11}S_{22} - S_{21}S_{12}|$ less than 1.

The stability factor is given by

$$K = \frac{1 - |S_{11}|^2 - |S_{22}|^2 + |S_{11}S_{22} - S_{21}S_{12}|^2}{|2S_{21}S_{12}|}$$

Since all of the expressions given in this section involve complex variable calculations, it is most practical to solve them with the aid of computer programs. Software is available not only to solve these equations, but also to simulate circuit response and optimize amplifier design. Touchstone, available from EEsof of Westlake Village, California, is an example of a powerful microwave/RF circuit simulation package. Touchstone will run on the IBM PC as well as on more powerful computers, such as the VAX series.

A design procedure for a narrowband, low-noise amplifier would be as follows:

1. Select an appropriate transistor.
2. Determine the class A bias conditions for minimum noise figure and design a bias network to achieve these conditions.
3. Determine the input impedance necessary for minimum noise figure from the optimum reflection coefficient.
4. Design a matching network to present this impedance to the transistor.
5. Determine the output impedance necessary for a conjugate match.
6. Design a matching network to present this impedance to the transistor.
7. Compute the transducer gain of the amplifier.
8. Ensure stability of the amplifier.

Example 14-1

Design a single-stage, small-signal, narrowband RF amplifier with a noise figure of less than 2.5 dB and a gain of approximately 10 dB at a frequency of 6 GHz. The amplifier is to have 50-ohm input and output impedances.

Solution

The device to be used in this amplifier is the Hewlett-Packard HFET-1101 GaAs FET. Measurements indicate that this device exhibits a minimum noise figure of 2.2 dB at 6 GHz when biased at $V_{GS} = -2$ volts, $V_{DS} = 3$ volts, and $I_{DS} = 10$ ma. The S-parameters of the HFET-1101 at these frequency and bias conditions are:

$$S_{11} = 0.674 \angle -152° \qquad S_{12} = 0.075 \angle 6.2°$$

$$S_{21} = 1.74 \angle 36.4° \qquad S_{22} = 0.60 \angle -92.6°$$

The optimum reflection coefficient is:

$$\Gamma_{s_{opt}} = 0.575 \angle 138°$$

Figure 14-2 illustrates a single supply bias arrangement for a GaAs FET. In this figure, the drain supply is fed to the transistor via a λ/4 section of microstrip line with one end shorted at RF. A λ/4 section of shorted transmission line has

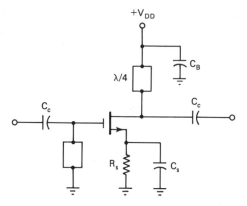

Figure 14-2 GaAs FET bias circuit.

an infinite RF input impedance. The line should be as narrow as practical to maximize its characteristic impedance. The gate is held at DC ground via another shorted microstrip line. The gate to source voltage is developed across R_s, which is bypassed by a microwave capacitor of 100 pF or greater. The drain supply should be bypassed by capacitors of approximately 100 pF, 0.01 μF and 1 μF in parallel, and the two coupling capacitors should be 1000 pF or greater.

The expression for the input reflection coefficient is

$$\Gamma_s = \frac{Z_s - Z_o}{Z_s + Z_o}$$

Solving this expression for Z_s,

$$Z_s = \frac{Z_o(1 + \Gamma_s)}{(1 - \Gamma_s)}$$

The source impedance corresponding to the optimum reflection coefficient of 0.575 $\angle 138° = -0.43 + j0.38$ is

$$Z_s = \frac{Z_o(1 + \Gamma_s)}{(1 - \Gamma_s)} = \frac{50(1 - 0.43 + j0.38)}{(1 + 0.43 - j0.38)} = 15.3 + j17.3$$

The input matching network must transform an impedance of $50 + j0$ to $15.3 + j17.3$ ohms. A simple microstrip matching network may be designed using a stub and a quarter wave impedance transformer. The stub will be in parallel with the transistor input so the optimum series impedance must be converted to a corresponding optimum admittance:

$$Y_{s_{opt}} = \frac{1}{Z_{s_{opt}}} = \frac{1}{15.3 + j17.3} = 0.028 - j0.032$$

A shorted microstrip line has an input impedance of

$$Z_i = jZ_o \tan \beta L$$

The admittance of such a line is

$$Y_i = \frac{1}{Z_i} = -j\frac{1}{Z_o \tan \beta L}$$

The susceptive portion of the optimum admittance may be realized by a shorted stub. The length of a stub with a characteristic impedance of 50 ohms is

$$\frac{1}{Z_o \tan \beta L} = 0.032$$

$$\tan \beta L = 0.63$$

$$\beta L = 0.56 \text{ radians}$$

$$\beta = \frac{2\pi}{\lambda}$$

$$L = \frac{\lambda(0.56)}{2\pi} = 0.09\lambda$$

The parallel conductance is equivalent to a resistance of

$$R = \frac{1}{G} = \frac{1}{0.028} = 35.7 \text{ ohms}$$

The 50-ohm source resistance may be transformed to 35.7 ohms by a quarter wave transformer of the following characteristic impedance:

$$Z_o = \sqrt{(50)(35.7)} = 42.3 \text{ ohms}$$

The input network then consists of a $\lambda/4$ length of 42.3-ohm microstrip line and a shorted shunt stub of length 0.09λ. The 200-ohm source resistor develops a gate to source voltage of -2 volts with a source current of 10 ma. For the desired drain to source voltage of 3 volts, the supply voltage must be $+5$ volts.

The reflection coefficient at the output is

$$\Gamma_o = S_{22} + \frac{S_{12}S_{21}\,\Gamma_s}{1 - \Gamma_s S_{11}}$$

Using a computer program to solve this expression, $\Gamma_o = -0.15 + j0.58$. The load impedance required for a conjugate match is:

$$Z_L = \frac{Z_o(1 + \Gamma_o^*)}{(1 - \Gamma_o^*)} = \frac{50(1 - 0.15 - j0.58)}{(1 + 0.15 + j0.58)} = 19.3 + j35.3$$

The output can also be matched with a shorted stub and quarter wave transformer. The load admittance is

$$Y_L = \frac{1}{Z_L} = \frac{1}{19.3 + j35.3} = 0.012 - j0.022$$

Using the same procedure as was used to design the input network, the length of a 50 ohm shorted stub is

$$L = 0.13\lambda$$

The quarter wave transformer has a characteristic impedance of

$$Z_o = \sqrt{(50)\left(\frac{1}{0.012}\right)} = 64.6 \text{ ohms}$$

The output network then consists of a microstrip stub of length 0.13λ, which must be shorted at RF but not at DC, and a $\lambda/4$ transformer with a characteristic impedance of 64.6 ohms.

The complete amplifier is illustrated in Figure 14-3.

As a final check, a computer program can be used to compute the transducer gain G_T and the conditions for stability:

$$G_T = 9.7 \text{ dB } (\approx 10 \text{ dB})$$

$$K = 1.28 \text{ } (>1)$$

$$|S_{11}S_{22} - S_{21}S_{12}|^2 = 0.39 \text{ } (<1)$$

Thus the amplifier is stable, exhibits a gain near the desired 10 dB, and has a noise figure less than 2.5 dB as desired.

Broadband low-noise amplifier design requires input and/or output networks to compensate for the variations in the S-parameter values that occur as a function of frequency. These networks must be designed by synthesis techniques, which

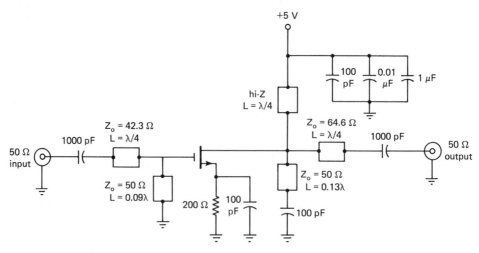

Figure 14-3 Low-noise 6 GHz RF amplifier.

are beyond the scope of this text, or with software packages such as E-Syn, discussed in Chapter 8.

Narrowband and broadband low-noise amplifiers are available prepackaged and may be considered a component in communications system designs. Manufacturers such as Avantek and Watkins-Johnson offer low-noise amplifier modules at frequencies up to 18 GHz and beyond. At VHF and below, low-noise amplifiers are available in integrated circuit form. An example of an integrated circuit low-noise amplifier is the Plessey SL560C, which operates from a single supply of 2 volts to 15 volts, and which is available in an 8-pin dual in-line package. The SL560C exhibits a noise figure less than 2 dB, gain up to 40 dB, and a bandwidth of 300 MHz.

14.2 INTERMEDIATE FREQUENCY AMPLIFIERS

In most communications receivers, the received signal is first amplified by a low-noise RF amplifier, which essentially establishes the noise figure of the system. The signal is then down-converted with a mixer to an intermediate frequency (IF), where the majority of the gain and selectivity of the receiver occurs. Some communications receivers utilize two or more IF frequencies. IF frequencies have become standardized over the years. Most AM broadcast receivers use 455 KHz as a standard, while FM broadcast receivers use 10.7 MHz. In non-broadcast applications, a number of different frequencies are used, but 70 MHz and 30 MHz are common. The choice of an IF frequency often involves a trade-off between selectivity and image frequency signal rejection. At low IF frequencies, filters with the desired sharp cutoff characteristics are normally easier to design. The concept of image rejection is best illustrated by an example. Assume an FM broadcast receiver with an IF of 10.7 MHz is tuned to a local station at 99.7 MHz. The local oscillator frequency would be $99.7 - 10.7 = 89.0$ MHz (or $99.7 + 10.7 = 110.4$ MHz). Assuming the local oscillator is 89.0 MHz, a signal at 78.3 MHz would also be mixed to the 10.7 MHz IF frequency. Since the undesired signal is $2 \times 10.7 = 21.4$ MHz away from the desired signal, filtering easily attenuates it to a negligible level. Receivers that use two IFs generally use a low frequency IF for selectivity and a high frequency IF for image rejection.

Selectivity in an IF amplifier is established by external filters. AM broadcast receivers typically use double-tuned transformers to achieve their selectivity. Crystal and ceramic filters are also often used in IF amplifiers to establish selectivity, especially at 10.7 MHz, where low-cost filters of this type are readily available.

IF amplifiers intended for FM systems often include limiter circuitry to remove amplitude noise prior to demodulation. Most IF amplifiers also have provisions for automatic gain control (AGC). In an AGC system, the level of the output signal is detected and used in a feedback control system to adjust the gain of the IF amplifier, so that variations in received signal level (which are slow compared to the modulation frequency) do not appear at the output of the amplifier.

Virtually all current communications receiver designs use integrated circuit IF amplifiers. Some integrated circuit IF amplifiers include additional functions on the same chip. An example of an integrated circuit IF amplifier is the Signetics NE604. The NE604 is available in a 16-pin dual in-line package and operates from a supply voltage of 4.5 volts to 8.0 volts. The chip contains two amplifiers that operate at frequencies up to 10.7 MHz, a signal strength indicator signal output, and a multiplier-type FM detector. The device has a gain of 30 dB and includes a mute input and limiter circuitry.

An example of an integrated circuit IF amplifier that may be used at higher frequencies is the Plessey SL610C. The SL610C is available in an 8-pin TO-5 package and operates from a supply voltage of 6 to 12 volts. The voltage gain of the SL610C is 20 dB with a cutoff frequency of 140 MHz. The device has an AGC range of 50 dB, but does not include limiting circuitry.

Another circuit of interest is the TDA7000 single-chip FM receiver manufactured by Signetics. The TDA7000 includes an RF amplifier, mixer, local oscillator, IF amplifier, limiter, demodulator, mute detector, and mute switch in a single 18-pin dual in-line package. The TDA7000 operates from a supply voltage from 2.7 volts to 10 volts and provides an audio output voltage of 1.3 volts.

14.3 POWER AMPLIFIERS

Power amplifiers are most often used in communications systems to increase the power level of a frequency generating circuit, such as an oscillator or a digital frequency synthesizer, to a value suitable for transmission. Power amplifiers may be operated class A or class B when linear operation is required (e.g., AM or SSB) or, most commonly, class C. A class C power amplifier is suitable for FM or digital modulation and exhibits a higher efficiency than does a class A or class B amplifier.

Power amplifiers are large-signal amplifiers, so the small-signal S-parameters used in low-noise amplifier design are not applicable. Semiconductor manufacturers supply input and output impedance data for their power devices under typical operating conditions in lieu of S-parameter data. The input and output networks for a power amplifier are normally designed to provide a conjugate match to the device impedances to maximize power gain.

Power amplifiers are available as components with integral bias networks and input and output impedance matching networks. Such devices normally are specified for 50-ohm input and output impedances; they thus may be easily cascaded for a desired gain or power level.

Power amplifier design is illustrated in this section by two examples. The first example illustrates a VHF power amplifier, which utilizes lumped components in its matching networks, and an L-band power amplifier, which utilizes microstrip techniques.

Example 14-2

A 50 MHz oscillator followed by two power amplifier stages provides an output power of 32 dBm at 50 ohms. Design an amplifier to increase this power output to a minimum of 10 watts at 50 ohms.

Solution

The required output power in dBm is:

$$P_o(\text{dBm}) = 10 \log \frac{P_o}{1.0 \text{ mw}} = 10 \log \frac{10}{10^{-3}} = 40 \text{ dBm}$$

The amplifier must provide at least 8 dB gain to bring the 32 dBm signal up to 40 dBm.

The transistor selected for this design is the Motorola MRF233 NPN silicon RF power transistor, which provides a gain of at least 10 dB and an output power of 15 watts at 50 MHz. The MRF233 is specified at a supply voltage of 12.5 volts, but may be operated from 8 volts to 16 volts as long as its maximum power dissipation limits are observed. The series equivalent input and output impedances of the MRF233 at 50 MHz are:

$$Z_i = 1.15 - j2.40$$

$$Z_o = 6.20 - j4.55$$

The matching networks discussed in Section 8.7 will be used in this example. Since the bandwidth was not specified, the Qs of the matching networks may be selected arbitrarily. In RF power amplifier design, the input network is normally used to set the amplifier bandwidth, and the output network is broadband to minimize the effects of changes in load VSWR. For this example, set $Q_{\text{in}} = 10$ and $Q_{\text{out}} = 5$. The network illustrated in Figure 14-4 will be used for both input and output matching.

The input network must match an impedance of $1.15 - j0.240$ into $50 + j0$ with a Q of 10. The design procedure has been outlined in Section 8.7.

1. $Q = 10$
2. $X_{L_1} = QR_s + X_{c_s} = 10 (1.15) - 2.40 = 9.1$

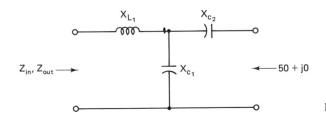

Figure 14-4   Matching network.

3. $A = \sqrt{\left[\dfrac{R_s\,(1+Q^2)}{R_L}\right] - 1} = \sqrt{\left[\dfrac{1.15\,(1+10^2)}{50}\right] - 1} = 1.15$

4. $B = R_s\,(1+Q^2) = 1.15\,(1+10^2) = 116.2$

5. $X_{c_2} = AR_L = (1.15)(50) = 57.5$

6. $X_{C_1} = \dfrac{B}{Q-A} = \dfrac{116.2}{10-1.15} = 13.1$

7. $L_1 = \dfrac{X_{L_1}}{2\pi f} = \dfrac{9.1}{(6.28)(50\times10^6)} = 29.0\times10^{-9}\,\text{H} = 29\ \text{nH}$

$C_1 = \dfrac{1}{2\pi f X_{C_1}} = \dfrac{1}{(6.28)(50\times10^6)(13.1)} = 243\times10^{-12} = 243\ \text{pF}$

Choose a standard value of 240 pF.

$C_2 = \dfrac{1}{2\pi f X_{C_2}} = \dfrac{1}{(6.28)(50\times10^6)(57.5)} = 55.4\times10^{-12} = 55.4\ \text{pF}$

Choose a standard value of 56 pF.

The output network must match an impedance of $6.20 - j4.55$ to $50 + j0$ with a Q of 5. The design procedure is as follows:

1. $Q = 5$

2. $X_{L_1} = QR_s + X_{c_s} = (5)(6.20) - 4.55 = 26.5$

3. $A = \dfrac{R_s(1+Q^2)}{R_L} - 1 = \dfrac{6.20\,(1+25)}{50} - 1 = 1.49$

4. $B = R_s\,(1+Q^2) = 6.20\,(1+25) = 161.2$

5. $X_{c_2} = AR_L = (1.49)(50) = 74.5\ \Omega$

6. $X_{c_1} = \dfrac{B}{Q-A} = \dfrac{161.2}{5-1.49} = 45.9\ \Omega$

7. $L_1 = \dfrac{X_{L_1}}{2\pi f} = \dfrac{26.5}{(6.28)(50\times10^6)} = 8.44\times10^{-8} = 84.4\ \text{nH}$

$C_1 = \dfrac{1}{2\pi f X_{C_1}} = \dfrac{1}{(6.28)(50\times10^6)(45.9)} = 69.4\times10^{-12} = 69.4\ \text{pF}$

Choose a standard value of 68 pF.

$C_2 = \dfrac{1}{2\pi f X_{C_2}} = \dfrac{1}{(6.28)(50\times10^6)(74.5)} = 42.7\times10^{-12} = 42.7\ \text{pF}$

Choose a standard value of 47 pF.

For a class C amplifier the base must be grounded at DC. This is normally accomplished with a radio frequency choke (RFC), which exhibits a high impedance

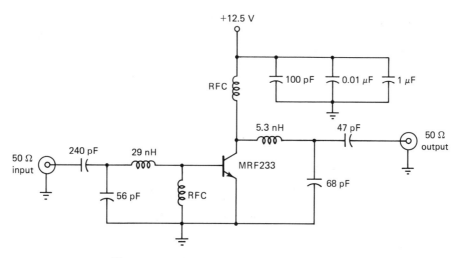

Figure 14-5   50 MHz RF power amplifier circuit.

at the RF center frequency (50 MHz) and a low DC resistance.   A 10 μH choke would be adequate in this application.   The collector also must be fed through an RFC; in addition, it is good practice to bypass the DC supply to the amplifier with three capacitors, one of approximately 1000 pF, one of 0.01 μF, and one of 1.0 μF to minimize the possibility of oscillation.   The completed circuit is illustrated in Figure 14-5.

Since the amplifier provides gain in excess of the desired 8 dB, it may be necessary to insert an attenuator between the driver and the amplifier input to achieve the desired 10 watts output.   Actual circuit losses, however, may negate the requirement for an attenuator.

Example 14-3

Design a power amplifier to increase the output of an L-band (1.26 GHz) FM transceiver from 1 watt to 10 watts.

Solution

The transistor selected must exhibit at least 10 dB gain at 1.26 GHz and be capable of continuous operation at an output power of 10 watts.

The Motorola MRF2005 RF power transistor is suitable for this application, since it exhibits a power gain of 12.5 dB at 1.3 GHz with a 28-volt supply in a common base configuration.   The device is capable of continuous operation at 13 watts output.   The input and output impedances of the device under the required operating conditions are

$$Z_i = 1.5 + j8.6$$

$$Z_o = 3.1 - j5.3$$

The input may be conjugate matched with a microstrip shunt stub and quarter wave impedance transformer. For a conjugate match, the matching network must transform $50 + j0$ ohms to the complex conjugate of the input impedance or

$$Z_{in}^* = 1.5 - j8.6$$

The admittance corresponding to this impedance is

$$Y_{in}^* = \frac{1}{Z_{in}^*} = \frac{1}{1.5 - j8.6} = \frac{1\angle 0°}{8.73 \ \angle -80.11} = 0.115\angle 80.11 = 0.02 + j0.113$$

The input impedance of an open circuit microstrip line is

$$Z_i = -jZ_o \cot \beta L$$

The admittance corresponding to this impedance is

$$Y_i = \frac{1}{Z_i} = j\frac{1}{Z_o \cot \beta L}$$

The length of an open circuit microstrip stub that realizes the desired suspectance may be calculated from

$$\frac{1}{Z_o \cot \beta L} = 0.113$$

For a characteristic impedance of 50 ohms,

$$\cot \beta L = \frac{1}{(50)(0.113)} = 0.18$$

$$\beta L = 1.39 \text{ radians}$$

$$\beta = \frac{2\pi}{\lambda}$$

$$L = \frac{1.39 \ \lambda}{2\pi} = 0.22\lambda$$

The parallel conductance is equivalent to a resistance of

$$R = \frac{1}{G} = \frac{1}{0.02} = 50 \text{ ohms}$$

In this special case, a quarter wave matching transformer is not necessary, since the resistance already equals 50 ohms. An arbitrary length of 50-ohm microstrip line would be necessary, however, to provide a path to the input connector.

The output may also be matched with a microstrip shunt stub and a quarter wave transformer. The network must transform $50 + j0$ ohms to the conjugate of the output impedance or

$$Z_{out}^* = 3.1 + j5.3$$

The admittance corresponding to this impedance is

$$Y^*_{out} = \frac{1}{Z^*_{out}} = \frac{1}{3.1 + j5.3} = \frac{1\angle 0°}{6.14 \angle 59.68} = 0.163\angle -59.68° = 0.08 - j0.14$$

The stub may be realized by a shorted microstrip line. The admittance of a shorted stub is

$$Y_i = -j\frac{1}{Z_o \tan \beta L}$$

The length of the stub may be calculated as follows:

$$\frac{1}{Z_o \tan \beta L} = 0.14$$

For a characteristic impedance of 50 ohms,

$$\tan \beta L = \frac{1}{(50)(0.14)} = 0.143$$

$$\beta L = 0.14 \text{ radians}$$

$$L = \frac{0.14\lambda}{2\pi} = 0.022\lambda$$

The parallel conductance is equivalent to a resistance of

$$R = \frac{1}{G} = \frac{1}{0.08} = 12.5 \text{ ohms}$$

This resistance may be transformed to 50 ohms with a quarter wave transformer with characteristic impedance given by

$$Z_o = \sqrt{(50)(12.5)} = 25 \text{ ohms}$$

This characteristic impedance would require a rather wide line with low dielectric constant substrates, so it might be desirable to cascade two impedance transformers if such a substrate is used.

The emitter may be grounded at DC with a narrow (high characteristic impedance) $\lambda/4$ microstrip line. The collector may also be fed with $+28$ volts DC with a narrow $\lambda/4$ microstrip line. The shorted microstrip stub in the output matching network should be placed at RF ground with a capacitor of approximately 500 pF (510 pF is a standard value). Coupling capacitors of approximately 1000 pF should be used at the input and output of the amplifier. The collector supply of the amplifier should be bypassed with capacitors of 100 pF, 0.01 μF and 1.0 μF.

The complete amplifier is illustrated in Figure 14-6. Ideally, this amplifier will provide power in excess of the desired 10 watts output for 1 watt input. Various circuit losses will reduce this figure somewhat. If it is necessary to limit the power

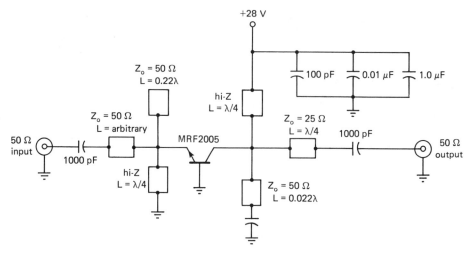

Figure 14-6 L-band RF power amplifier circuit.

output to exactly 10 watts, an attenuator may be inserted at the input to the amplifier. The need for an attenuator is best determined by actual measurements.

RF amplifiers are available as components with a 50-ohm input and output impedance. These amplifier modules require a single power supply. An example of an RF amplifier module is the TRW MX20, which provides a gain of 20 dB and a power output of 20 watts over the 400–512 MHz frequency range. The MX20 operates from a nominal power supply voltage of + 12.5 volts.

Another example of an RF power module is the Motorola MHW820-3, which provides a gain of 17 dB and a power output of 18 watts over the 870–950 MHz band. The MHW820-3 has 50-ohm input and output impedances and operates from a nominal 12.5-volt supply.

RF power amplifiers are available as components in modular form at frequencies as high as 18 GHz and as low as HF. Most RF power amplifier modules utilize thin film hybrid construction techniques.

EXERCISES

1. A GaAs FET has the following S-parameters at 2 GHz:

$$S_{11} = 0.5\underline{/-100°} \qquad S_{12} = 0.05\underline{/10°}$$

$$S_{21} = 2.0\underline{/45°} \qquad S_{22} = 0.6\underline{/-80°}$$

The optimum reflection coefficient for the device is:

$$\Gamma_{S_{opt}} = 0.6 \angle 150°$$

The minimum noise figure (1.5 dB) occurs at a bias point of $V_{GS} = -3$ volts, $V_{DS} = 5$ volts and $I_{DS} = 10$ ma. Design a single stage, small signal, narrowband, low noise amplifier with this device, assuming 50-ohm input and output impedances.

2. Calculate the transducer gain for the amplifier of Exercise 1.
3. Calculate the stability conditions for the amplifier of Exercise 1.
4. A bipolar power transistor has the following input and output impedances in a common base configuration at 2 GHz when operated class C with a 28-volt supply:

$$Z_{in} = 2 + j10 \qquad Z_{out} = 4 - j6$$

Design a 2 GHz power amplifier using the above device.

15

RF Transmission Lines

All communications systems utilize transmission lines. Transmitters and receivers may use them internally to direct RF energy between points, as well as externally to direct RF energy to or from an antenna or another piece of equipment. Transmission lines used in RF communications systems include flexible or semi-rigid coaxial cable, waveguide and microstrip. The type of transmission line used in a system depends upon factors such as the frequency of operation, the power level, the allowable attenuation, and various mechanical and environmental considerations.

Since the general characteristics of transmission lines have been discussed previously, this chapter will emphasize practical considerations.

15.1 COAXIAL TRANSMISSION LINES

A coaxial cable is a transmission line in which one conductor completely surrounds the other, separated by a dielectric material. Figure 15-1 illustrates a coaxial cable. The center conductor is either a solid wire or several wires twisted together. Copper wire is normally used, but sometimes it is coated with a highly conductive material, such as silver. The dielectric material is usually polystyrene, polyethylene or polytetrafluoroethylene (PTFE or teflon). The outer conductor is a woven wire braid in the case of flexible coaxial cables, or a solid tube in the case of semi-rigid cables. Flexible cables and some semi-rigid cables also have a vinyl insulating and/ or protective jacket surrounding the outer conductor.

Coaxial cables are characterized by their distributed parameters. The capacitance per unit length of a coaxial cable is given by

$$C = \frac{24.1 \; \varepsilon_r'}{\log \left(\dfrac{d_2}{d_1} \right)}$$

In this expression, C is the distributed capacitance in pF per meter, d_1 is the diameter

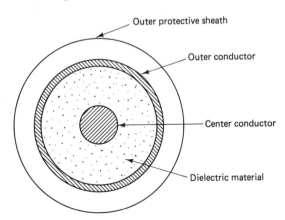

Figure 15-1 Coaxial cable.

of the center conductor, d_2 is the inside diameter of the outer conductor, and ε_r' is the real part of the complex dielectric constant of the material between the conductors (*not* the effective dielectric constant used in microstrip calculations!).

The inductance per unit length of a coaxial cable is given by

$$L = 0.461 \log \left(\frac{d_2}{d_1} \right)$$

where L is the distributed inductance in μH per meter.

The series resistance per unit length of a copper conductor coaxial cable is

$$R = 4.14 \times 10^{-6} \sqrt{f} \left(\frac{1}{d_2} + \frac{1}{d_1} \right)$$

In this expression, R is the distributed series resistance in ohms per meter, and f is the frequency in Hz.

The parallel conductance of a coaxial line is given by

$$G = 2\pi \times 10^{-12} fC \frac{\varepsilon_r''}{\varepsilon_r'}$$

In this expression, G is the conductance in Siemens per meter, C is the distributed capacitance in pF per meter and ε_r'' is the imaginary part of the complex dielectric constant.

Example 15-1

A coaxial cable is constructed with a PTFE (teflon) dielectric, a copper center conductor with a diameter of 3 mm, and a copper outer conductor whose inside diameter is 10 mm.   Compute the distributed parameters of this cable at 1 GHz.

Solution

The dielectric constant of PTFE at 1 GHz is approximately $2.1 - j0.003$.

$$d_1 = 3 \times 10^{-3} \text{ m}$$

$$d_2 = 1 \times 10^{-2} \text{ m}$$

$$f = 1 \times 10^9 \text{ Hz}$$

$$\varepsilon_r' = 2.1$$

$$\varepsilon_r'' = 3 \times 10^{-3}$$

$$C = \frac{24.1 \, \varepsilon_r'}{\log \left(\dfrac{d_2}{d_1} \right)} = \frac{(24.1)(2.1)}{\log \left(\dfrac{10}{3} \right)} = 97 \text{ pF/m}$$

$$L = 0.461 \log \left(\frac{d_2}{d_1} \right) = 0.461 \log \left(\frac{10}{3} \right) = 0.24 \ \mu H/m$$

$$R = 4.14 \times 10^{-6} \sqrt{f} \left(\frac{1}{d_2} + \frac{1}{d_1} \right) = (4.14 \times 10^{-6})(3.16 \times 10^4)(4.3 \times 10^2)$$

$$= 56.3 \ ohm/m$$

$$G = 2\pi \times 10^{-12} fC \frac{\varepsilon_r''}{\varepsilon_r'} = \frac{(2)(3.14)(10^{-12})(10^9)(97)(0.003)}{2.1}$$

$$= 8.7 \times 10^{-4} \ Siemens/m$$

The characteristic impedance of a coaxial line may be calculated from its distributed parameters as discussed in Section 11.1:

$$Z_o = \sqrt{\frac{R + j\omega L}{G + j\omega C}}$$

For low-loss coaxial lines (the usual case), the characteristic impedance may be computed from

$$Z_o = \frac{138}{\sqrt{\varepsilon_r'}} \log \left(\frac{d_2}{d_1} \right)$$

In both of the above expressions for characteristic impedance, Z_o is in ohms.

Example 15-2

Compute the characteristic impedance of the coaxial cable of Example 15-1.

Solution

The low-loss approximation is appropriate for this case:

$$Z_o = \frac{138}{\sqrt{\varepsilon_r'}} \log \left(\frac{d_2}{d_1} \right) = \frac{138}{\sqrt{2.1}} \log \left(\frac{10}{3} \right) = 50.1 \ ohms$$

Most communications systems are designed for a characteristic impedance of 50 ohms, although a few use either 75 ohms or 93 ohms.

The velocity of propagation of an electromagnetic wave in a coaxial cable is normally expressed as a percentage of the speed of light (3×10^8 m/sec). The percentage is computed as follows:

$$\text{percentage} = \frac{100}{\sqrt{\varepsilon_r'}}$$

Example 15-3

Compute the velocity of propagation (v_p) in the coaxial cable of Example 15-1.

Solution

$$\text{percentage} = \frac{100}{\sqrt{\varepsilon_r'}} = \frac{100}{\sqrt{2.1}} = 69\%$$

$$v_p = (0.69)(3 \times 10^8) = 2.07 \times 10^8 \text{ m/sec}$$

Attenuation in a coaxial transmission line is the result of conductor losses and dielectric losses. An approximate expression for conductor loss in a copper coaxial cable is

$$\alpha_c = \frac{1.8 \times 10^{-7} \sqrt{f}\left(\dfrac{1}{d_2} + \dfrac{1}{d_1}\right)}{Z_o}$$

In this expression, α_c is the attenuation due to the conductor resistance loss expressed in dB per meter. An approximate expression for the dielectric loss in a coaxial cable is

$$\alpha_d = 9.1 \times 10^{-8} \sqrt{\varepsilon_r'}\, f \tan \delta$$

where α_d is the attenuation due to the dielectric loss expressed in dB per meter, and $\tan \delta$ is the loss tangent given by

$$\tan \delta = \frac{\varepsilon_r''}{\varepsilon_r'}$$

The total attenuation is the sum of these two factors:

$$\alpha = \alpha_d + \alpha_c$$

If the dielectric constant is approximately constant with frequency, it is apparent that conductor losses are proportional to the square root of frequency, while the dielectric losses are directly proportional to frequency. Thus, as the frequency is increased, the total cable attenuation increases with the dielectric losses becoming the dominant factor at higher frequencies.

Example 15-4

Calculate the attenuation of the cable in Example 15-1 at 1.0 GHz.

Solution

$$d_1 = 3 \times 10^{-3} \text{ m}$$

$$d_2 = 1 \times 10^{-2} \text{ m}$$

$$f = 1 \times 10^9 \text{ Hz}$$

$$Z_o = 50.1 \text{ ohms}$$

$$\varepsilon_r' = 2.1$$

$$\varepsilon_r'' = 3 \times 10^{-3}$$

$$\alpha_c = \frac{1.8 \times 10^{-7} \sqrt{f}\left(\dfrac{1}{d_2} + \dfrac{1}{d_1}\right)}{Z_o}$$

$$= \frac{(1.8 \times 10^{-7})(3.16 \times 10^4)(3.3 \times 10^2 + 1 \times 10^2)}{50.1}$$

$$\alpha_c = 4.9 \times 10^{-2} \text{ dB/m}$$

$$\alpha_d = 9.1 \times 10^{-8} \sqrt{\varepsilon_r'} \, f \tan \delta$$

$$= (9.1 \times 10^{-8})(1.45)(1 \times 10^9)\frac{3 \times 10^{-3}}{2.1}$$

$$\alpha_d = 2.0 \times 10^{-1} \text{ dB/m}$$

$$\alpha = \alpha_c + \alpha_d = 4.9 \times 10^{-2} + 2.0 \times 10^{-1} = 2.5 \times 10^{-1} \text{ dB/m}$$

A coaxial line is normally operated at a frequency where the electric field lines in the cable extend radially from the center conductor to the outer conductor and the magnetic field lines make concentric circles around the center conductor. Since the electric field lines and the magnetic field lines are both perpendicular to the direction of propagation, the wave is said to be propagating down the line in a transverse electromagnetic (TEM) mode. A cutoff frequency exists for each type of coaxial cable above which higher order modes with different electric and magnetic field configurations will also exist. Operation at higher order modes is generally avoided in coaxial cables due to the potential increases in VSWR and attenuation. The cutoff frequency of a coaxial line may be calculated from

$$f_c = \frac{3 \times 10^8}{\pi \varepsilon_r' \left(\dfrac{d_2 + d_1}{2}\right)}$$

In this expression, f_c is in Hz; d_2 and d_1 are in meters.

Example 15-5

Calculate the cutoff frequency of the cable in Example 15-1.

Solution

$$d_1 = 3 \times 10^{-3} \text{ m}$$

$$d_2 = 1 \times 10^{-2} \text{ m}$$

$$\varepsilon_r' = 2.1$$

$$f_c = \cfrac{3 \times 10^8}{\pi \varepsilon_r' \left(\cfrac{d_2 + d_1}{2} \right)} = \frac{3 \times 10^8}{(3.14)(1.45)(6.5 \times 10^{-3})}$$

$$= 10.1 \times 10^9 = 10.1 \text{ GHz}$$

The continuous wave (CW) power rating of a coaxial cable is also of considerable importance. It depends upon such factors as frequency, temperature, altitude, and VSWR. As each of these factors increases, the power that can be safely handled by the cable decreases. The CW power rating for a given type of cable is available on the manufacturer's data sheets for that cable.

Flexible coaxial cables are normally identified by a military designator (RG number). An example of a popular 50-ohm flexible cable is RG-214 which has an outer jacket diameter of 11 mm, a dielectric inside diameter of 7 mm, and a center conductor diameter of 2 mm. RG-214 exhibits a capacitance of 96.8 pF per meter and has 0.3 dB per meter attenuation at 1 GHz. It has a CW power rating of 190 watts at 1 GHz but only 37 watts at 10 GHz.

Semi-rigid coaxial cables are normally identified by their outside diameter in inches. An example of a popular semi-rigid coaxial cable is SRC-141, which has an outside diameter of 0.141 inches or 3.58 mm. SRC-141 exhibits 0.38 dB per meter attenuation at 1 GHz and can handle 600 watts of continuous power at that frequency. Semi-rigid coaxial cable suffers less attenuation and can handle more power than the equivalent size in a flexible cable. Semi-rigid cables are essentially 100% shielded, whereas flexible cables with a wire braid outer conductor may only provide 90% shielding. The disadvantage of semi-rigid cable is that it cannot be flexed without damage and generally may only be bent into a desired shape once in its lifetime.

An important consideration when using coaxial cables is the type of connector used. UHF connectors may be used up to approximately 500 MHz, BNC-type connectors may be used up to 4 GHz (with low VSWR), and type-N connectors may be used up to 11 GHz. Military numbers are often used to specify coaxial connectors. An example of a UHF connector is the PL-259 plug, which is appropriate for coaxial cables, such as RG-214. A BNC plug suitable for RG-214 would be the UG-959, and the type-N plug for the same cable would be the UG-21. Small-diameter semi-rigid coaxial cables normally use SMA-type coaxial connectors. Other connectors used in the microwave range include the APC7 and APC3.5 series of connectors.

15.2 WAVEGUIDE TRANSMISSION LINES

In general, any transmission line can be considered a waveguide in that it guides electromagnetic energy from one point to another. In practice, however, the term waveguide is usually reserved for a class of transmission lines constructed of hollow pipe of rectangular, circular, or occasionally elliptical cross section. Waveguides are normally only used at microwave frequencies due to size constraints. They are easier to fabricate than coaxial cables and can be designed to exhibit less attenuation than coaxial cables. Their main disadvantage is their general lack of mechanical flexibility, although flexible (but lossy) waveguides are available.

Figure 15-2 illustrates a section of rectangular waveguide. The mode of operation of the waveguide is determined by the *a* dimension in Figure 15-2. The *a* dimension must be at least one-half wavelength at the operating frequency. The *b* dimension in Figure 15-2 determines the attenuation and power handling capability of the waveguide; *b* is normally chosen to be approximately one-half of *a*.

Although an infinite number of modes of operation, each with its own unique electric and magnetic field distribution, are possible in a waveguide, the guide is normally operated in the so-called dominant mode. The dominant mode is the mode with the lowest cutoff frequency.

In general, two classes of modes may exist in a waveguide—the so-called transverse electric (TE) mode or transverse magnetic (TM) mode. A TE wave in a waveguide has a magnetic field component parallel to the axis of the waveguide and an electric field that is everywhere transverse to the axis. A TM wave, on the other hand, has an electric field component parallel to the axis and a magnetic field that is everywhere transverse to the axis.

In addition to the TE or TM designation for modes of operation in a rectangular waveguide, numerical subscripts are used. The subscripts are integers that indicate the number of half-period variations in field intensity along the *a* dimension (m) and *b* dimension (n) of the waveguide respectively. A given mode would be described as a TE_{mn} or TM_{mn} mode. The dominant mode in a rectangular waveguide is the TE_{10} mode, which indicates that the wave is of the transverse electric class with a single half wave variation of field intensity along the *a* dimension and no variation in field intensity along the *b* dimension. The lower cutoff frequency

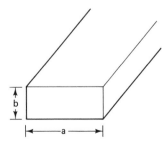

Figure 15-2   Rectangular waveguide.

for a mode may be calculated from

$$f_c = 1.5 \times 10^8 \sqrt{\left(\frac{m}{a}\right)^2 + \left(\frac{n}{b}\right)^2}$$

In this expression, f_c is the lower cutoff frequency in Hz, a and b are the waveguide dimensions in meters, and m and n are the integers indicating the mode.

Example 15-6

A rectangular waveguide is 3.7 cm by 1.8 cm (inside dimensions). Calculate the cutoff frequency of the dominant mode.

Solution

The dominant mode in a rectangular waveguide is the TE_{10} mode, so $m = 1$ and $n = 0$.

$$f_c = 1.5 \times 10^8 \sqrt{\left(\frac{m}{a}\right)^2 + \left(\frac{n}{b}\right)^2}$$

$$= 1.5 \times 10^8 \sqrt{\left(\frac{1}{0.037}\right)^2 + \left(\frac{0}{0.018}\right)^2}$$

$$= 4.1 \times 10^9 = 4.1\,\text{GHz}$$

Example 15-7

Calculate the cutoff frequency and identify the mode closest to the dominant mode for the waveguide in Example 15-6.

Solution

TM modes with $m = 0$ or $n = 0$ are not possible in a rectangular waveguide. The TE_{01}, T_{20} and TE_{02} modes are possible, however. The cutoff frequencies for these modes are:

TE_{01}: 8.3 GHz TE_{20}: 8.1 GHz TE_{02}: 16.7 GHz

Thus the TE_{20} mode has the lowest cutoff frequency of all modes except the dominant TE_{10} mode.

The waveguide could be used over the frequency range of 4.1 GHz to 8.1 GHz in the dominant mode. In practice, the recommended range of operation for a waveguide having these dimensions would be somewhat less, to provide a margin for manufacturing tolerances and other sources of uncertainty.

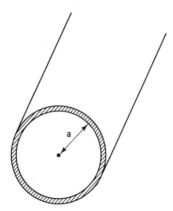

Figure 15-3 Circular waveguide.

Figure 15-3 illustrates a section of circular waveguide. The mode of operation of a circular waveguide is determined by its radius a. Modes are identified as TE_{mn} or TM_{mn}, where m is the total number of full period variations of either field component along a circular path concentric with the wall, and n indicates the number of half wave variations across a diameter.

The dominant mode in a circular waveguide is the TE_{11} mode, and the mode with the next lowest cutoff frequency is the TM_{01} mode. The cutoff frequency for the TE_{11} mode is given by

$$f_c = \frac{5.52 \times 10^8}{2\pi a}$$

The cutoff frequency for the TM_{01} mode is given by

$$f_c = \frac{7.22 \times 10^8}{2\pi a}$$

In both of these expressions, f_c is in Hz and a is in meters.

Example 15-8

Calculate the operating bandwidth of a circular waveguide with a diameter of 2 cm.

Solution

For the dominant (TE_{11}) mode:

$$f_c = \frac{5.52 \times 10^8}{2\pi a} = \frac{5.52 \times 10^8}{(6.28)(1 \times 10^{-2})} = 8.8 \times 10^9 = 8.8 \text{ GHz}$$

For the TM_{01} mode:

$$f_c = \frac{7.22 \times 10^8}{2\pi a} = \frac{7.22 \times 10^8}{(6.28)(1 \times 10^{-2})} = 1.15 \times 10^{10} = 11.5 \text{ GHz}$$

The maximum bandwidth of this waveguide is from 8.8 GHz to 11.5 GHz (a total of 2.7 GHz).

A number of standard waveguides are designated by Electronic Industries Association (EIA) WR numbers or military RG numbers. An example of standard waveguide is WR-137 or RG-50/U, which is specified for operation from 5.85 GHz to 8.20 GHz, and which exhibits an attenuation of 0.075 dB per meter to 0.094 dB per meter over its operating range, and has a power rating of over 500,000 watts. The military rectangular waveguides have a circular outside diameter and a rectangular bore. The waveguide characteristics in this case, of course, are determined by the dimensions of the rectangular bore.

Waveguides may be interfaced with coaxial lines by terminating the outer conductor of the coax on the waveguide wall and allowing the center conductor to extend through a hole in the waveguide parallel to the electric field (axis of a coaxial line is normal to *a* dimension in a rectangular waveguide). The center conductor acts as an antenna, radiating energy that then propagates down the guide. Normally the energy flow is desired in a single direction, so the end of the waveguide near the "antenna probe" is shorted. To minimize the VSWR on the coaxial line, the probe must be matched to the coax's characteristic impedance. The probe impedance is a function of its length and distance from the shorted end of the waveguide.

15.3 MICROSTRIP TRANSMISSION LINES

Microstrip is one of the most versatile components available to the communications system designer. Microstrip techniques were discussed in Section 11.2, and applications were covered in Chapters 12 and 14. Coaxial cable and waveguide are used both internally and externally in transmitters and receivers, while microstrip transmission lines are almost always used internally. The primary use of microstrip lines is the realization of impedances for matching active devices and the construction of impedance transformers via the use of quarter-wavelength line segments. Microstrip is also used internally to route RF energy from subsystems, such as amplifiers or filters, to coaxial connectors and sometimes to route signals between subsystems. At frequencies where microstrip transmission lines are practical, small-diameter semi-rigid coaxial cable is frequently used, and SMA type connectors are available to provide a transition from the coax to the microstrip. Microstrip transmission lines are also used to feed microstrip antennas, but usually only short runs with a quick transition to an SMA connector are used.

The primary reason that microstrip is not used for long runs is that it is inherently not shielded. This permits line radiation and increased loss. Despite this drawback, in applications where its use is appropriate, microstrip is a most valuable RF component.

EXERCISES ━━━━━━━━━━━━━━━━━━━━━━━━━━━━━━━━━━

1. A PTFE dielectric coaxial cable has a copper center conductor with a diameter of 2 mm and a copper outer conductor whose inside diameter is 10 mm. Compute the distributed parameters and characteristic impedance of this cable at 700 MHz.
2. Calculate the attenuation of the cable in Exercise 1 at 700 MHz.
3. Calculate the cutoff frequency of the cable in Exercise 1.
4. A non-standard rectangular waveguide is fabricated from copper and has inside dimensions of 10 cm × 5 cm. Calculate the cutoff frequency of this waveguide.
5. A section of circular pipe with an inside diameter of 5 cm is to be used as a waveguide. Calculate the operating bandwidth of this guide.

16

Antennas

An antenna may be considered as an impedance matching device between a transmission line and free space.   Antennas are essential components in many types of communications systems.   All antennas exhibit common characteristics, such as input impedance, gain, beamwidth, and polarization.   Although there are numerous types of antennas in use, this chapter will discuss only those of primary importance, including the dipole, the Yagi-Uda array, the helix, the horn, the reflector antenna, and the microstrip antenna.

Antennas are used for both receiving and transmitting, and most exhibit a property known as reciprocity—they have identical receiving and transmitting characteristics.   All of the antennas discussed in this chapter exhibit reciprocity.

Since antenna theory is a mature subject, most textbooks treat it in a highly mathematical manner, which tends to obscure the fundamental principles and design techniques.   The approach taken in this chapter is to emphasize understanding and applications, utilizing only appropriate mathematics.

16.1 ANTENNA FUNDAMENTALS

An *isotropic antenna* is a fictitious device that radiates (and hence receives—by reciprocity) equally well in all directions.   Antenna parameters, such as gain and directivity, are often quoted with respect to an isotropic source (antenna).   Another concept that is important to the study of antennas is the propagation of electromagnetic waves; Maxwell's equations, which form the basis of electromagnetic field theory, require a time-varying electric field perpendicular to a time-varying magnetic field for an electromagnetic wave propagating in a direction mutually perpendicular to both fields.   The electric field, magnetic field, and direction of propagation form a right-hand coordinate system.   Antennas radiate spherical waves, but at large distances away from the antenna, the spherical waves may be approximated over a small region by plane waves.   Plane waves, although fictitious, simplify antenna analysis and design considerably.

One of the most useful characteristics of an antenna is its field patterns.   An antenna pattern is measured by applying RF power to the antenna to be measured while moving a receiving probe in a constant radius circle about the antenna to measure electric field strength or relative power.   To completely describe the pattern of an antenna, an infinite number of such measurements would be needed for each possible circle.   Fortunately, antennas may be adequately described by measuring only two patterns and, in addition, if the probe is located sufficiently far from the antenna under measurement (far-field), the plane wave assumption may be used.   Antenna patterns, therefore, normally consist of measurements in the so-called two principal planes, the E-plane (the plane formed by the direction of the electric field and the direction of propagation), and the H-plane (the plane formed by the direction of the magnetic field and the direction of propagation).

The relative power is normally expressed in dB and normalized to 0 dB at the main lobe maximum.   The sidelobes are usually described by their level relative

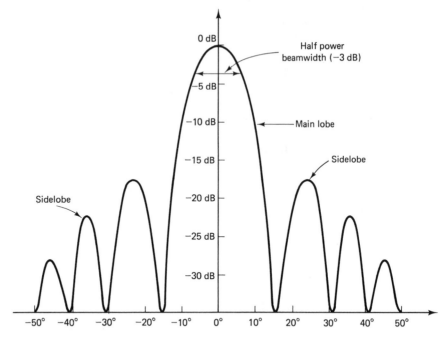

Figure 16-1 Antenna pattern.

to the main lobe maximum. In a typical antenna pattern, Figure 16-1, the first pair of sidelobes is down by approximately 17.5 dB compared to the peak of the main lobe.

The half-power beamwidth of an antenna is defined by the angular distance between the main lobe's −3 dB points. In Figure 16-1, this is approximately 20°. The minimum far-field distance at which pattern measurements may be made is approximated from $R \geq 2d^2/\lambda_o$, where R is the radial distance from the antenna under test to the measurement point, d is the maximum linear dimension of the antenna, and λ_o is the free-space wavelength.

Example 16-1

A 10 GHz antenna with a maximum linear dimension of 10 cm is to be used as a test antenna on an antenna range. Calculate the minimum separation between this antenna and the receiving probe antenna if far-field measurements are desired.

Solution

$$d = 0.1 \text{ m}$$

$$\lambda_o = \frac{c}{f} = \frac{3 \times 10^8}{10 \times 10^9} = 0.03 \text{ m}$$

$$R \geq \frac{(2)(0.1)^2}{0.03} = 0.67 \text{ m}$$

The directivity of an antenna is a measure of the concentration of energy in the main lobe and may be approximated by

$$D = \frac{4\pi}{\theta_1 \theta_2}$$

In the above expression, D is the directivity ratio, and θ_1 and θ_2 are the 3 dB beamwidths, in radians, of the two principal plane patterns. Directivity is normally expressed in dB and is with respect to an isotropic antenna, or Directivity = 10log D.

Example 16-2

Antenna range measurements on an antenna at 6 GHz reveal an E-plane beamwidth of 15° and an H-plane beamwidth of 20°. Estimate the directivity of this antenna.

Solution

$$\theta_1 = 15° = 0.26 \text{ radians}$$

$$\theta_2 = 20° = 0.35 \text{ radians}$$

$$D = \frac{4\pi}{\theta_1 \theta_2} = \frac{(4)(3.14)}{(0.26)(0.35)} = 138.0 = 21.4 \text{ dB}$$

The gain of an antenna is a parameter closely related to directivity:

$$G = \eta D$$

In this expression, G is the gain of the antenna, η is the ohmic efficiency of the antenna, and D is the antenna's directivity. A real antenna has less than 100% ($\eta = 1$) efficiency, due to conductivity losses and mismatch losses. Another type of antenna "efficiency" relates to the electrical effective aperture area of an antenna compared to its physical aperture area (see Section 16.5).

Example 16-3

A 1 GHz satellite antenna has an E-plane beamwidth of 12° and an H-plane beamwidth of 10°. The antenna's conductivity and mismatch losses total 3 dB. Estimate the gain of this antenna.

Solution

$$\theta_1 = 12° = 0.21 \text{ radians}$$

$$\theta_2 = 10° = 0.17 \text{ radians}$$

$$dB = 10 \log \frac{P_o}{P_i}$$

$$-3 = 10 \log \frac{P_o}{P_i}$$

$$\frac{P_o}{P_i} = \eta = 0.5$$

$$D = \frac{4\pi}{\theta_1 \theta_2} = \frac{(4)(3.14)}{(0.21)(0.17)} = 351.8$$

$$G = \eta D = (0.5)(351.8) = 175.9 = 22.5 \text{ dB}$$

Note that the gain also could have been calculated by subtracting the loss in dB from the directivity in dB.

The pattern of Figure 16-1 has only one main lobe and is often called a pencil-beam pattern. Some antennas are designed to have many equal-amplitude lobes. The parameters of such antennas may not be estimated from the approximations given in this section, and techniques beyond the scope of this text must be used to treat them.

The polarization of an antenna is the direction of its electric field. Polarization may be linear or elliptical. Linear polarizations are called horizontal polarization or vertical polarization, with a reference such as the earth's surface. In elliptical polarization, the direction of the electric field rotates in either a clockwise (CW) or counter-clockwise (CCW) direction at a rate given by the radian frequency of the wave. The ratio of the major axis power of the ellipse to its minor axis power is called the axial ratio. The most commonly used elliptical polarization is circular polarization, which has an axial ratio of 1.0. Clockwise rotation as viewed from the rear of the transmitting antenna is called right-hand circular polarization, and counterclockwise rotation is called left-hand circular polarization. For maximum energy transfer, a receiving antenna must have the same polarization as the transmitting antenna. In the ideal case, cross-polarized antennas (horizontal to vertical or right-hand circular to left-hand circular) transfer no energy. One measure of an antenna's performance is its cross-polarization ratio, which is the ratio of the desired polarization component to the undesired polarization component in dB.

Example 16-4

A 6 GHz vertically polarized antenna is fed with RF power at one end of an antenna range, and a receiving probe is connected to a power meter at the other end of the range.   The power meter reads 10 mW when the probe is vertical and 100 μW when the probe is horizontal.   Assuming the probe's depolarization is negligible, calculate the depolarization ratio of the antenna under test.

Solution

$$\text{Depolarization ratio} = 10 \log \frac{P_v}{P_H} = 10 \log \frac{10 \times 10^{-3}}{100 \times 10^{-6}} = 20 \text{ dB}$$

Since an antenna may be considered as an impedance matching device between a transmission line and free space, it must exhibit some value of input impedance itself.   The antenna input impedance must equal the transmission line impedance if the VSWR on the line is to be minimized.   Many coaxial-fed antenna and transmission line systems are designed with a nominal input impedance value of $50 + j0$ ohms.   If the antenna exhibits an input impedance other than the transmission line impedance, some type of matching device is normally utilized. Input impedance is a unique property of each type of antenna and thus will be discussed individually for each antenna type covered in the remaining sections of this chapter.

16.2 DIPOLE ANTENNAS

The dipole is perhaps the most basic antenna used in communications.   It consists of a conductor, usually at least one-half wavelength long, split at the center and fed with a transmission line.   Figure 16-2 illustrates a dipole antenna.   In Figure 16-2, the length of the dipole is l, so each segment of the antenna is $l/2$.   The current distribution along a dipole antenna depends upon its length.   Figure 16-3 illustrates current distributions for four common dipole lengths.

In Figure 16-3, note that the current distribution depends upon the electrical length of the dipole (length with respect to wavelength, λ_o) and is sinusoidal in shape.   Note also that in each case, the current falls to zero at the ends of the dipole, since an open circuit exists at those points.

The current distribution in a dipole antenna determines its pattern, so the

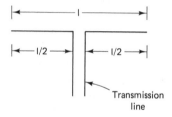

Transmission
line   **Figure 16-2** Dipole antenna.

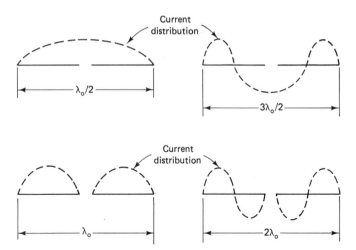

Figure 16-3 Dipole current distributions.

pattern is a function of electrical length. The normalized pattern for the widely used half-wave dipole ($l = \lambda_o/2$) is given by

$$E(\theta) = \frac{\cos\left(\dfrac{\pi}{2}\cos\theta\right)}{\sin\theta}$$

Figure 16-4 is a sketch of this antenna pattern. Note that a half-wave dipole exhibits two maxima (main lobes) broadside to the antenna. The E-plane and H-plane patterns have the same shape for this dipole. The expression for the pattern of the half-wave dipole and Figure 16-4 represent the shape of the electric or magnetic field distribution in the far field. The 3 dB beamwidth (half-power beamwidth) can be found by setting the pattern expression equal to 0.707, finding the four angles that satisfy the equation and subtracting to determine the half-power beamwidth. This exercise results in a beamwidth of 78 degrees for the half-wave dipole. The gain of a half-wave dipole is 2.14 dB over an isotropic antenna.

A full-wave dipole ($l = \lambda_o$) also exhibits two broadside lobes, but the beamwidth narrows to 47 degrees. As the length of a dipole is further increased, additional lobes appear. A 3/2 wavelength dipole, for example, has six lobes.

The input impedance of a dipole antenna involves calculations beyond the scope of this textbook. These calculations indicate that the impedance of a half-wave dipole is approximately $73 + j42$ ohms. If the length of the half-wave dipole is reduced to approximately 95% of a half wavelength, the inductive portion of the input impedance goes to zero, and the dipole is said to be resonant. The input impedance under these conditions is approximately $67 + j0$ ohms.

It is often desirable to feed a dipole antenna with a coaxial transmission line. A coaxial cable is inherently an unbalanced transmission line (as opposed to a

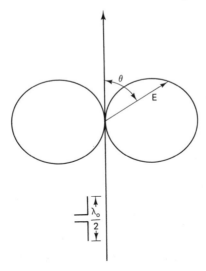

Figure 16-4 Antenna pattern of $\lambda_o/2$ dipole.

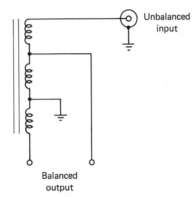

Unbalanced
input

Balanced
output

Figure 16-5 Balun.

balanced transmission line, where the two conductors have the same capacitance to ground and the same series inductance per unit length).   Unbalanced coaxial lines directly feeding a balanced antenna, such as a dipole, will have RF current flowing on the outside of the coax that radiates and may distort the antenna pattern. The solution to this problem is the use of a device called a balun.   A balun (*bal*anced to *un*balanced), often implemented as windings on a ferrite core, will provide the necessary transition to prevent line radiation.   Figure 16-5 illustrates a balun.

Example 16-5

A resonant dipole antenna is fed with a 50-ohm coaxial cable through a 1:1 balun. Calculate the VSWR on the transmission line.

Solution

$$Z_{in} = 67 + j0 \text{ ohms}$$

The reflection coefficient is

$$\Gamma = \frac{Z_L - Z_o}{Z_L + Z_o} = \frac{67 - 50}{67 + 50} = 0.15$$

The VSWR on the transmission line is

$$\text{VSWR} = \frac{1 + |\Gamma|}{1 - |\Gamma|} = \frac{1 + 0.15}{1 - 0.15} = 1.35$$

The 1:1 balun provides the necessary balanced to unbalanced transformation but without transforming impedance levels.

An antenna configuration closely related to the half-wave dipole is the quarter-wave vertical antenna. In this configuration, one-half of a half-wave dipole is erected vertically over a ground plane, usually formed by radial wires buried underground for AM broadcast band antennas or the body of a car or aircraft for vehicular antennas. The ground plane creates an image of the missing half of the dipole and performance similar to a half-wave dipole is obtained. The input impedance of a quarter-wave vertical is somewhat lower than that of the dipole—approximately 50 ohms, making it a good match to a 50-ohm cable.

Single dipole antenna are used primarily at frequencies less than 30 MHz, but are often also used at higher frequencies as the driven element in array or reflector antennas.

16.3 ARRAY ANTENNAS

As discussed in Section 16.2, the basic dipole antenna has a bidirectional pattern with a rather wide beamwidth. In most communications systems, information must be transferred between two or more fixed points, so it is desirable to use an antenna with a unidirectional pattern and a narrow beamwidth. The narrow beamwidth implies greater directivity and thus greater antenna gain. If an antenna with high gain is available for use in a communications system, other parts of the system may be simplified. For example, a 100-watt transmitter feeding a 0 dB gain antenna is equivalent to a 1-watt transmitter feeding a 20 dB gain antenna. A 1-watt transmitter is much more compact, lighter and less costly than a 100-watt unit and consumes less primary power.

The directivity of the basic dipole antenna may be increased by creating an array of two or more dipoles spaced so that they concentrate the radiation in a given desired direction. If there is negligible coupling between the dipoles, the receiving point is sufficiently distant from the transmission point, and the fields arriving from each dipole are in phase, equal current magnitudes in the dipoles will result in a power gain approximately proportional to the number of dipole elements.   An array whose elements all receive power through a transmission line is known as a driven array.   An array with one driven element and additional elements electromagnetically coupled to the driven element (these elements are called parasitic elements) is known as a Yagi-Uda array.

Antenna elements in an array may be arranged either in a parallel or a collinear (end-to-end) configuration.   The elements may also be arranged horizontally or vertically, depending upon the direction of linear polarization desired. Circularly polarized arrays are designed using both horizontal and vertical elements on the same boom with two driven elements.   For circular polarization, the power must be split between the two driven elements and 90 degrees of phase shift must be inserted in the feed to one of the driven elements if they both lie in the same plane.   The direction of the phase shift in the driven element shifted in phase determines the sense of the circular polarization (RHC or LHC).   An array may radiate perpendicular to the array axis (broadside array) or parallel to the array axis (endfire array), depending upon the relative phasing of the currents in the elements.

The most popular type of array antenna is the Yagi-Uda array.   This antenna requires direct connections only to a single driven element.   The amplitude and phase of the currents electromagnetically induced in the other elements depend upon the spacing between the driven element and the so-called "parasitic" elements as well as upon the "tuning" (length) of the parasitic elements.

Since a unidirectional antenna is most desirable for point-to-point communications, array antennas are normally designed so that at least one parasitic element is adjusted to cancel radiation from the backside of the array.   An element so adjusted is called a reflector element.   A figure of merit for such an antenna is its front-to-back ratio.   This ratio measures the capability to reduce the antenna's response to interfering signals arriving from directions opposite the desired direction.   A parasitic element adjusted so that it enhances radiation in the forward direction is called a director element.   Practical array antennas have at least one reflector and one or more directors.   The spacing between the reflector and the driven element is normally set to approximately $0.2 \, \lambda_o$, which is a compromise between maximum forward gain and maximum front-to-back ratio.   The first director in a Yagi-Uda array is spaced approximately $0.15 \, \lambda_o$ from the driven element, the second approximately $0.2 \, \lambda_o$ from the first, and the third through fifth approximately $0.3 \, \lambda_o$ from the previous director.   Subsequent directors, if used, should be spaced at approximately $0.4 \, \lambda_o$ apart.   The directivity of a Yagi-Uda array as a function of its total number of elements is approximately $D = 0.62n + 8.5$ dB, where n is the total number of elements in the array, including the driven element.

The input impedance of a Yagi-Uda array depends upon the length of the driven element as well as upon the element spacing.   Since a typical figure is 25

to 30 ohms, a matching network or transformer must be used to match this impedance to the 50-ohm coaxial cables typically used in communications systems.

The bandwidth of a Yagi-Uda array is a function of its element spacing and
the thickness of the conductors used for elements. Bandwidth in an antenna is
determined by changes (with frequency) in input impedance which affect the VSWR
on the feed line. Typical bandwidths of a Yagi-Uda array for a 2:1 VSWR are
4%.

Example 16-6

Design a 500-MHz Yagi-Uda array antenna for an E-plane and H-plane beamwidth
of 31 degrees.

Solution

The relationship between directivity and beamwidth is

$$D = \frac{4\pi}{\theta_1\theta_2}$$

Assuming $\theta_1 = \theta_2 = 31$ degrees, the corresponding directivity is

$$D = \frac{(4)(3.14)}{(0.54)(0.54)} = 43.1 = 16.3 \text{ dB}$$

The number of elements required for this directivity may be found from:

$$D = 0.62n + 8.5$$
$$16.3 = 0.62n + 8.5$$
$$n = 12.6 \rightarrow 13$$

A thirteen-element array is required with a driven element, a reflector, and
11 directors.

The length of the driven element should be approximately 0.49 λ_o or 29.4
cm, the length of the reflector approximately 0.51 λ_o or 30.6 cm, and the length
of the directors should be about 0.45 λ_o or 27.0 cm. The spacing between the
driven element and reflector should be 0.2 λ_o or 12.0 cm. The first director should
be 0.15 λ_o or 9.0 cm from the driven element, the second director should be 0.2
λ_o or 12.0 cm from the first, the third through fifth should be spaced at 0.3 λ_o or
18.0 cm, and the remaining directors should be spaced at 0.4 λ_o, or 24.0 cm. Figure
16-6 illustrates the antenna design. This antenna would have a bandwidth of
approximately 20 MHz centered at 500 MHz.

The design procedures given in this section are approximate, and more detailed methods (often computer-aided) for designing optimum Yagi-Uda array
antennas are available in the literature. For most purposes, the approximate
procedures are adequate, since it is difficult to account quantitatively for factors
such as height above ground, ground conductivity, the effects of nearby structures

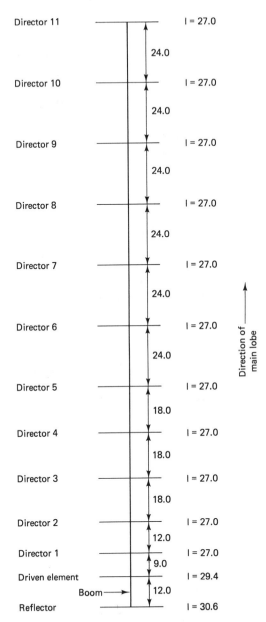

Director 11   l = 27.0

                    24.0

Director 10   l = 27.0

                    24.0

Director 9   l = 27.0

                    24.0

Director 8   l = 27.0

                    24.0

Director 7   l = 27.0

                    24.0

Director 6   l = 27.0

                    24.0

Director 5   l = 27.0

                    18.0

Director 4   l = 27.0

                    18.0

Director 3   l = 27.0

                    18.0

Director 2   l = 27.0

                    12.0

Director 1   l = 27.0

                    9.0

Driven element   l = 29.4

Boom   12.0

Reflector   l = 30.6

Direction of main lobe

Figure 16-6 Yagi-Uda array design of Example 16-6.(Note; all dimensions in cm.)

such as towers, buildings, trees, and the like.   Yagi-Uda arrays are primarily used from approximately 10 MHz to 1 GHz, due to construction considerations.

Array antennas composed of all driven elements are not limited to dipole elements and may be constructed from any antenna type.   The primary requirement for such arrays is some technique for controlling the amplitude and phase of the RF power supplied to each antenna.

16.4 HELIX ANTENNAS

The helix (or helical) antenna is extremely easy to construct and adjust for proper operation and thus is quite popular. It has satisfactory dimensions from approximately 100 MHz to 10 GHz. In addition to its appeal because of its simplicity, the helix features circular polarization. A sketch of a helix antenna is given in Figure 16-7. In this figure, α is the pitch angle of the helix, S is the spacing between turns, D is the diameter, A is the axial length, and d is the diameter of the helix conductor. Other helix parameters include the circumference C and the number of turns n. These parameters are related as follows:

$$C = \pi D$$

$$\alpha = \tan^{-1} \frac{S}{\pi D}$$

$$A = nS$$

It is common practice to refer to helix dimensions in terms of wavelength. As an example, C_λ is the helix circumference measured in free-space wavelengths.

Example 16-7

A 6-turn 1 GHz helix antenna is 0.5 meter in length and has a diameter of 10 cm. Calculate its pitch angle, the spacing between turns and its circumference and diameter in wavelengths.

Solution

$$s = \frac{A}{n} = \frac{0.5}{6} = 0.083 \text{ m}$$

$$\alpha = \tan^{-1} \frac{s}{\pi D} = \tan^{-1} \frac{0.083}{(3.14)(0.1)} = 14.8°$$

$$\lambda_o = \frac{3 \times 10^8}{f} = \frac{3 \times 10^8}{1 \times 10^9} = 0.3 \text{ m}$$

$$D_\lambda = \frac{D}{\lambda_o} = \frac{0.1}{0.3} = 0.33$$

$$C_\lambda = \pi D_\lambda = (3.14)(0.33) = 1.04$$

Depending upon its dimensions (with respect to a wavelength), a helix may radiate in either the normal mode or in the axial mode. In the normal mode, the

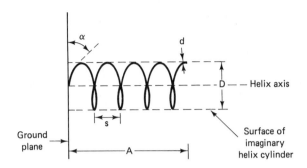

Figure 16-7 Helix antenna.

main lobe is perpendicular to the axis of the helix, while in the axial mode, the main lobe is along the axis. Almost all helix antennas utilize the axial mode, so the discussion in this section will be limited to that mode of operation.

The helix antenna operates in the axial mode when its circumference, C_λ, is of the order of 1 wavelength. The normal limits on the axial mode are:

$$\frac{3}{4} \leq C_\lambda \leq \frac{4}{3}$$

The input impedance of a helix antenna operating in the axial mode is a relatively constant function of frequency if the number of turns is greater than 3 and the pitch angle is between 12 and 15 degrees. Under these conditions the input impedance is essentially resistive and is given by $R = 140\ C_\lambda$ ohms.

Example 16-8

Compute the input impedance of the helix antenna in Example 16-7.

Solution

$$R = 140\ C_\lambda = (140)(1.04) = 145.6 \text{ ohms}$$

Since the input impedance of a helix operating in the axial mode ranges from 105 ohms to 187 ohms, an impedance transformation must be made to match commonly used coaxial lines.

The half-power beamwidth in degrees of an axial mode helix is

$$BW = \frac{52}{C_\lambda \sqrt{nS_\lambda}}$$

The directivity of an axial mode helix is

$$D = 15\ C_\lambda^2\ nS_\lambda \quad (D = \text{directivity})$$

The axial ratio of a helix operating in the axial mode is given by

$$AR = \frac{2n + 1}{2n}$$

In this expression as the number of turns is increased, the polarization approaches pure circular polarization ($AR = 1.0$).

Example 16-9

Design a 2.4 GHz helix antenna with a directivity of 15 dB. Compute the beamwidth, axial ratio and input impedance of the antenna.

Solution

Choose $C_\lambda = 1$ and $\alpha = 14°$:

$$\lambda_o = \frac{3 \times 10^8}{2.4 \times 10^9} = 0.13 \text{ m}$$

$$D = \frac{C}{\pi} = \frac{0.13}{3.14} = 0.041 \text{ m} \ (D = \text{diameter})$$

$$S = \pi D \tan \alpha = (3.14)(0.041)(0.249) = 0.032 \text{ m}$$

$$S_\lambda = \frac{S}{\lambda_o} = \frac{0.032}{0.13} = 0.246$$

$$D_{db} = 10 \log D \ (D = \text{directivity})$$

$$D = \log^{-1} \frac{D_{dB}}{10} = \log^{-1} \frac{15}{10} = 31.6 \ (D = \text{directivity})$$

$$n = \frac{D}{15 C_\lambda^2 S_\lambda} = \frac{31.6}{(15)(1)^2(0.246)} = 8.56 \ (D = \text{directivity})$$

With $n = 9$, the directivity is

$$D = 15 \, C_\lambda^2 n S_\lambda = (15)(1)^2(9)(0.246) = 33.2 = 15.2 \text{ dB} \ (D = \text{directivity})$$

$$A = nS = (9)(0.032) = 0.29 \text{ m}$$

$$BW = \frac{52}{C_\lambda \sqrt{n S_\lambda}} = \frac{52}{1 \sqrt{(9)(0.246)}} = 35°$$

$$AR = \frac{2n + 1}{2n} = \frac{(2)(9) + 1}{(2)(9)} = 1.06$$

$$R = 140 \, C_\lambda = (140)(1) = 140 \text{ ohms}$$

Summary

$\alpha = 14°$ $D = 15.2$ dB (directivity)

$D = 4.1$ cm (diameter) $BW = 35°$

$S = 3.2$ cm $AR = 1.06$

$n = 9$ $R = 140$ ohms

$A = 29$ cm

The ground plane in a helix antenna is perpendicular to the antenna at the feed end, is usually circular, and should have a diameter of at least $\lambda_o/2$.

16.5 HORN ANTENNAS

The horn antenna is widely used in communications for terrestrial point-to-point applications, satellite earth coverage satellite antennas, laboratory standards, and as a feed for large reflector antennas. A horn antenna provides a gradual transition between a waveguide and free space. Horns may be fed with coaxial cable by using a short section of shorted waveguide at the horn input with a probe adjusted to match the cable impedance. Figure 16-8 illustrates the most common horn types.

The conical horn and the pyramidal horn have E-plane and H-plane beamwidths that are approximately equal, while the two sectoral horns produce "fan beams," which are much narrower in one plane than in the other.

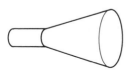

Conical horn

E-plane sectoral horn

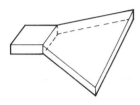

H-plane sectoral horn

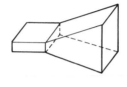

Pyramidal horn

Figure 16-8 Common horn types.

A horn antenna is a member of a class of antennas known as aperture antennas. The directivity and beamwidth of an aperture antenna depend upon the aperture dimensions and the field distribution over the aperture. For a given aperture, directivity is maximum and the beamwidth is narrowest for a uniformly illuminated aperture. Since a horn antenna is driven by a waveguide, the field distribution, and hence the illumination, is a function of the mode of operation of the waveguide. Waveguides are normally operated in their dominant mode, which is the TE_{10} mode for rectangular waveguide and the TE_{11} mode for circular waveguide. The electric field distribution (and the illumination) provided by a rectangular waveguide operating in the TE_{10} mode is constant across the b dimension (y-direction) and a half-cosine distribution that tapers to zero at the edge of the waveguide across the a dimension (x-direction). The field distribution in a circular waveguide operating in the TE_{11} mode is maximum at the center of the waveguide and tapers to zero at the edges.

Approximate expressions for the beamwidths of pyramidal and sectoral horns are

$$BW_E \approx 55\,\frac{\lambda_o}{l_y} \qquad BW_H \approx 70\,\frac{\lambda_o}{l_x}$$

In the above expressions, BW_E and BW_H are the E-plane and H-plane beamwidths in degrees respectively, λ_o is the free-space wavelength in meters, l_y is the aperture dimension in the y-direction in meters (b dimension of waveguide) and l_x is the aperture dimension in the x-direction in meters (a dimension of waveguide).

The beamwidths of a conical horn are approximately

$$BW_E \approx 30\,\frac{\lambda_o}{a} \qquad BW_H \approx 35\,\frac{\lambda_o}{a}$$

In the above expression, BW_E and BW_H are the E-plane and H-plane beamwidths in degrees respectively, λ_o is the free-space wavelength in meters, and a is the radius of the circular aperture in meters.

The directivity of a uniformly illuminated aperture is given by

$$D = \frac{4\pi A_p}{\lambda_o^2}$$

In this expression, D is the directivity, A_p is the physical area of the aperture and λ_o is the free-space wavelength. Horn antennas do not have uniformly illuminated apertures, so it is common practice to introduce a parameter called the aperture efficiency, η_a, to account for non-uniform illumination. The aperture efficiency for pyramidal and sectoral horns is approximately 0.6, and the aperture efficiency of conical horns is about the same. The expression for the directivity of a non-uniformly illuminated aperture, such as that of a horn antenna, is thus given by

$$D = \frac{4\pi\eta_a A_p}{\lambda_o^2}$$

η_a is a dimensionless constant between 0 and 1. The aperture efficiency is distinct from the antenna ohmic efficiency, η, discussed in Section 16.1, which depends upon resistive and sometimes mismatch losses. For a horn antenna the ohmic losses are relatively low and $\eta \approx 1$.

Example 16-10

A horn antenna with $l_x = 17.5$ cm and $l_y = 11.9$ cm is to be used as a receiving antenna at 10 GHz. Calculate its gain and beamwidth at this frequency.

Solution

$$G \approx D = \frac{4\pi\eta_a A_p}{\lambda_o^2}$$

$$\eta_a = 0.6$$

$$A_p = l_x l_y = (0.175)(0.119) = 0.021 \text{ m}^2$$

$$\lambda_o = \frac{3 \times 10^8}{10 \times 10^9} = 3 \times 10^{-2} \text{ m}$$

$$G = \frac{(4)(3.14)(0.6)(0.021)}{(3 \times 10^{-2})^2} = 1.8 \times 10^2 = 22.6 \text{ dB}$$

$$BW_E = 55\frac{\lambda_o}{l_y} = \frac{(55)(3 \times 10^{-2})}{11.9 \times 10^{-2}} = 13.9°$$

$$BW_H = 70\frac{\lambda_o}{l_x} = \frac{(70)(3 \times 10^{-2})}{(17.5 \times 10^{-2})} = 12.0°$$

Example 16-11

Calculate the diameter and beamwidths of a conical horn that is to have a gain of 20 dB at 5 GHz.

Solution

$$G_{dB} = 10 \log G$$

$$G = \log^{-1}\left(\frac{G_{dB}}{10}\right) = \log^{-1}(2) = 100$$

$$G = D = \frac{4\pi\eta_a A_p}{\lambda_o^2}$$

$$\eta_a = 0.6$$

$$A_p = \pi a^2$$

$$\lambda_o = \frac{3 \times 10^8}{5 \times 10^9} = 6 \times 10^{-2} \text{ m}$$

$$a = \sqrt{\frac{G\lambda_o^2}{4\pi^2\eta_a}} = \sqrt{\frac{(100)(6 \times 10^{-2})^2}{(4)(3.14)^2(0.6)}} = 0.12 \text{ m}$$

$$d = 2r = (2)(0.12) = 0.24 \text{ m}$$

$$BW_E = 30\,\frac{\lambda_o}{a} = \frac{(30)(6 \times 10^{-2})}{0.12} = 15°$$

$$BW_H = 35\,\frac{\lambda_o}{a} = \frac{(35)(6 \times 10^{-2})}{0.12} = 17.5°$$

16.6 REFLECTOR ANTENNAS

Reflector antennas are widely used in communications systems primarily because of their potential for high gain with a relatively simple mechanical structure. The Yagi-Uda array antenna with a single reflector element and a driven element is the most simple form of a reflector antenna. Other types of reflectors include rectangular flat sheets, corner reflectors, circular reflectors, elliptical reflectors, hyperbolic reflectors, and the most common form of reflector, the parabolic reflector. The discussion in this section will emphasize antennas with parabolic reflectors.

Figure 16-9 illustrates the geometrical properties of the parabolic reflector. If an isotropic source is placed at the focus of a parabolic reflector, all energy reflected from the parabolic surface will form a plane wave front (parallel rays) with equal phase at and beyond the aperture plane.

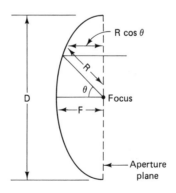

Figure 16-9 Geometrical properties of parabolic reflector.

This property implies that $2F = R + R \cos \theta$ or $R = 2F/(1 + \cos \theta)$. The above equation describes the required surface contour for a parabolic reflector. Parabolic reflectors are often described in terms of their F/D ratio (see Figure 16-9). A "deep" reflector has a smaller F/D ratio than a "shallow" reflector; practical values of F/D range from approximately 0.3 to 2.0.

Parabolic antennas usually utilize a unidirectional feed to minimize direct radiation from the feed. A small horn antenna is frequently used in this application. The antenna pattern of the horn illuminates the aperture formed by the parabolic reflector in a non-uniform manner. The illumination of the aperture determines its directivity, beamwidth and sidelobe level.

The directivity of a parabolic reflector is

$$D = \frac{4\pi\eta_a A_p}{\lambda_o^2}$$

The above expression is identical to the one given in the previous section for the horn antenna.

For the parabolic reflector antenna, however, the aperture efficiency η_a consists of several terms. η_a is the product of an amplitude taper efficiency (due to the horn illumination pattern), a spillover efficiency, a phase error efficiency, and a cross-polarization efficiency. Additionally, aperture blockage by the feed horn and its supporting structure is normally included in the overall aperture efficiency. These factors combine to give an overall aperture efficiency of approximately 0.5 to 0.6, depending upon the illumination pattern by the feed.

The half-power beamwidth of a parabolic reflector antenna is approximately

$$BW = \frac{67\lambda_o}{D}$$

In this expression, BW is in degrees and λ_o and D are in meters.

In a parabolic reflector antenna fed with a small horn, the sidelobes are typically down approximately 27 dB from the main beam maximum. Uniform illumination would result in maximizing the directivity at the expense of increased sidelobe levels. Tapered illumination, as provided by a horn feed, reduces the sidelobe level but results in decreased directivity. The sidelobe levels can be reduced to the point where they are negligible by carefully controlled tapered illumination, but the resulting low directivity is normally unacceptable.

Example 16-12

A parabolic reflector with an F/D ratio of 0.4 is fed by a small horn placed at the focal point that is 60 cm from the vertex. Calculate the gain and beamwidth of this antenna at 10 GHz.

Solution

$$F/D = 0.4$$

$$F = 0.6 \text{ m}$$

$$D = \frac{F}{0.4} = \frac{0.6}{0.4} = 1.5 \text{ m}$$

$$A_p = \pi\left(\frac{D}{2}\right)^2 = (3.14)\left(\frac{1.5}{2}\right)^2 = 1.77 \text{ m}^2$$

$$\lambda_o = \frac{3 \times 10^8}{10 \times 10^9} = 0.03 \text{ m}$$

Assume $\eta_a = 0.5$ and $\eta = 1.0$.

$$G = D = \frac{4\pi\eta_a A_p}{\lambda_o^2} = \frac{(4)(3.14)(0.5)(1.77)}{(0.03)^2} = 12.4 \times 10^3 = 40.9 \text{ dB}$$

$$BW = \frac{67\lambda_o}{D} = \frac{(67)(0.03)}{1.5} = 1.3°$$

16.7 MICROSTRIP ANTENNAS

The microstrip antenna is one of the most versatile tools available to the communications system designer. The first practical microstrip antennas were developed in the early 1970s. A microstrip antenna in its simplest configuration consists of a radiating patch (fed with a microstrip transmission line or a coaxial connector) on one side of a dielectric substrate, and a ground plane on the other side. The patch may be any shape, but simple shapes, such as the rectangle and circle, are normally used to simplify the design and analysis. Figure 16-10 illustrates a rectangular microstrip antenna. Figure 16-11 illustrates a circular or disk microstrip antenna.

Advantages of microstrip antennas, in addition to their extreme simplicity, include light weight, thinness, low volume, and low profile; they are thus ideal for aerospace vehicles and satellites. In addition, microstrip antennas are low in cost and allow RF circuits to be fabricated on the same substrate as the antenna. The microstrip antenna is capable of linear or circular polarization with a simple change in feed position. The major disadvantages of microstrip antennas are their narrow bandwidth (1% to 5%), their relatively low gain compared to other antenna types, and their lower power handling capability. In many applications, however, the advantages far outweigh the disadvantages and the use of this antenna type is rapidly increasing.

Since the microstrip antenna has air dielectric above it ($\varepsilon_r = 1$), an effective dielectric constant must be computed for use in design equations. The effective

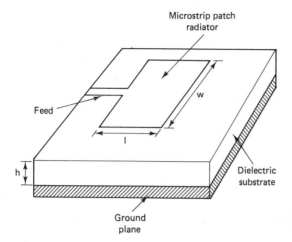

Figure 16-10   Rectangular microstrip antenna.

dielectric constant is computed as in Section 11.2 for $w/h > 1$ and is given by

$$\varepsilon_r' = \frac{\varepsilon_r + 1}{2} + \frac{\varepsilon_r - 1}{2}\left(1 + \frac{12h}{w}\right)^{-\frac{1}{2}}$$

The primary radiation mechanism of a rectangular microstrip antenna is the fringe field along the width of the patch. This fringe field effectively lengthens the dimension l by $2\Delta l$, where

$$\Delta l = \frac{0.412h(\varepsilon_r' + 0.3)\left(\dfrac{w}{h} + 0.264\right)}{(\varepsilon_r' - 0.258)\left(\dfrac{w}{h} + 0.8\right)}$$

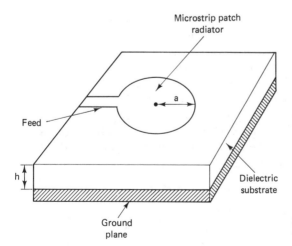

Figure 16-11   Circular microstrip antenna.

The initial step in the design of a rectangular microstrip antenna is selection of a suitable dielectric of appropriate thickness h. Substrates used in microstrip antennas have dielectric constants ranging from approximately 2.0 to 10.0. Normally dielectric materials with an $\varepsilon_r \approx 2$ are used in microstrip antennas to maximize the gain and bandwidth.

The width of a rectangular microstrip antenna is not particularly critical, but small widths result in inefficient antennas, and large widths result in the excitation of high order modes, which distort the antenna pattern. A practical width may be computed from

$$W = \frac{c}{2f}\left(\frac{\varepsilon_r + 1}{2}\right)^{-\frac{1}{2}}$$

In this expression, W is the width of the patch, c is the velocity of light, f is the design frequency, and ε_r is the substrate relative dielectric constant (not the effective dielectric constant ε_r').

The length of the rectangular microstrip antenna is quite critical and should be calculated from

$$l = \frac{c}{2f\sqrt{\varepsilon_r'}} - 2\Delta l$$

In this expression, l is the length of the patch, c is the velocity of light, f is the design frequency, ε_r' is the effective dielectric constant, and Δl is the length extension created by the fringe field.

Approximate expressions for the directivity of a microstrip antenna are

$$D \approx 6.6 \qquad (w < \lambda_o)$$

$$D \approx \frac{8.0w}{\lambda_o} \qquad (w \geq \lambda_o)$$

where D is the directivity, w is the width of the microstrip patch, and λ_o is the free-space wavelength. To compute the gain of the antenna, the directivity must be multiplied by the antenna efficiency. Efficiency in a microstrip antenna is a function of the dielectric constant and the thickness of the dielectric compared to the free-space wavelength. As an example, at a frequency of 1 GHz ($\lambda_o = 30$ cm) the efficiency ranges from approximately 90% for a dielectric thickness of 0.318 cm to approximately 50% for a thickness of 0.08 cm for a PTFE dielectric with $\varepsilon_r = 2.55$. For a 0.159 cm PTFE dielectric, the efficiency varies from approximately 10% to 98% as the frequency is varied from 100 MHz to 10 GHz.

The beamwidths of a rectangular microstrip antenna are given by

$$BW_H = 2\cos^{-1}\left[\frac{1}{2\left(1.0 + \frac{k_o w}{2}\right)}\right]^{\frac{1}{2}}$$

$$BW_E = 2 \cos^{-1} \left[\frac{7.03}{3k_o^2 l^2 + k_o^2 h^2} \right]^{\frac{1}{2}}$$

In these expressions, BW_H and BW_E are the H-plane and E-plane half-power beamwidths in degrees respectively, k_o is the free-space wavenumber given by $k_o = 2\pi/\lambda_o$, l is the length of the patch, and h is the dielectric thickness.

Example 16-13

Design a rectangular microstrip patch antenna for a frequency of 1.3 GHz and calculate the approximate gain and beamwidths.

Solution

Choose a 0.318 cm PTFE substrate with $\varepsilon_r = 2.55$.

$$w = \frac{c}{2f} \left(\frac{\varepsilon_r + 1}{2} \right)^{-\frac{1}{2}} = \frac{3 \times 10^8}{(2)(1.3 \times 10^9)} \left(\frac{2.55 + 1}{2} \right)^{-\frac{1}{2}}$$

$$= 0.087 \text{ m} = 8.7 \text{ cm}$$

$$\varepsilon_r' = \frac{\varepsilon_r + 1}{2} + \frac{\varepsilon_r - 1}{2} \left(1 + \frac{12h}{w} \right)^{-\frac{1}{2}}$$

$$= \frac{2.55 + 1}{2} + \frac{2.55 - 1}{2} \left(1 + \frac{(12)(3.18 \times 10^{-3})}{8.7 \times 10^{-2}} \right)^{-\frac{1}{2}} = 2.42$$

$$\Delta l = \frac{0.412h(\varepsilon_r' + 0.3)\left(\dfrac{w}{h} + 0.264 \right)}{(\varepsilon_r' - 0.258) \left(\dfrac{w}{h} + 0.8 \right)}$$

$$= \frac{(0.412)(3.18 \times 10^{-3})(2.72)(27.36)}{(2.16)(28.16)} = 2 \times 10^{-3} \text{ m}$$

$$l = \frac{c}{2f\sqrt{\varepsilon_r'}} - 2\Delta l$$

$$= \frac{3 \times 10^8}{(2)(1.3 \times 10^9)(2.42)^{\frac{1}{2}}} - (2)(2 \times 10^{-3})$$

$$= 7 \times 10^{-2} \text{ m} = 7 \text{ cm}$$

Since $w < \lambda_o$, $D \approx 6.6$.   Assume $\eta \approx 0.9$.

$$G = \eta D = (0.9)(6.6) = 5.94 = 7.4 \text{ dB}$$

$$BW_H = 2\cos^{-1}\left[\frac{1}{2\left(1 + \dfrac{k_o w}{2}\right)}\right]^{\frac{1}{2}}$$

$$k_o = \frac{2\pi}{\lambda_o} = \frac{(2)(3.14)}{23 \times 10^{-2}} = 27.3$$

$$BW_H = 2\cos^{-1}\left[\frac{1}{2\left\{1 + \dfrac{(27.3)(8.6 \times 10^{-2})}{2}\right\}}\right]^{\frac{1}{2}} = 122.7°$$

$$BW_E = 2\cos^{-1}\left[\frac{7.03}{3k_o^2 l^2 + k_o^2 h^2}\right]^{\frac{1}{2}}$$

$$= 2\cos^{-1}\left[\frac{7.03}{(3)(27.3)^2(7 \times 10^{-2})^2 + (27.3)^2(3.18 \times 10^{-3})^2}\right]^{\frac{1}{2}}$$

$$= 73.6°$$

Summary

$f = 1.3$ GHz	$G = 7.4$ dB
$h = 0.318$ cm	$BW_H = 122.7°$
PTFE	$BW_E = 73.6°$
$w = 8.7$ cm	$l = 7$ cm

Rectangular microstrip antennas may be fed either with a microstrip line or directly with a coaxial connector. The input impedance of a microstrip antenna will vary as the feed is moved from the center of the patch width toward the edge. The impedance at the center of the patch width exceeds 50 ohms, so a matching transformer section of microstrip line would be necessary to transform this impedance to the normally desired value of 50 ohms. A common procedure is to experimentally move the feed position toward the edge of the patch until a 50-ohm match is obtained. Coaxial connector feeds are also usually empirically adjusted for the best match.

The radius of a circular microstrip antenna is approximately

$$a = \frac{1.84c}{2\pi f \sqrt{\varepsilon_r}}$$

In the above expression, a is the radius of the patch in meters, c is the speed of light, f is the design frequency, and ε_r is the substrate relative dielectric constant.

Example 16-14

Design a circular microstrip antenna for a frequency of 1.3 GHz.

Solution

Choose a 0.318 cm PTFE substrate with $\varepsilon_r = 2.55$.

$$a = \frac{(1.84)(3 \times 10^8)}{(2)(3.14)(1.3 \times 10^9)(2.55)^{\frac{1}{2}}} = 4.2 \times 10^{-2}\,\text{m} = 4.2\,\text{cm}$$

A circular microstrip antenna has approximately the same gain and efficiency as a rectangular microstrip antenna on the same substrate. The beamwidths of a circular microstrip antenna are approximately 90% of the E-plane beamwidth and 65% of the H-plane beamwidth of a rectangular microstrip antenna on the same substrate. The circular antenna consumes slightly less area on the substrate, and its bandwidth is approximately 50% of the bandwidth of the rectangular antenna. The input impedance of a circular microstrip antenna is several hundred ohms, so an impedance transformer will normally be required to step this impedance down to 50 ohms.

The low gain and wide beamwidth of single patch microstrip antennas may be circumvented by using arrays of microstrip elements. The entire array and its feeds may be etched on a single substrate. Gains of up to 20 dB are possible with microstrip array antennas.

EXERCISES ━━━━━━━━━━━━━━━━━━━━━━━━━━

1. Calculate the far-field distance for an antenna with a maximum linear dimension of 50 cm operating at a frequency of 4 GHz.
2. Estimate the directivity in dB of a pencil beam antenna whose E-plane beamwidth is 25° and H-plane beamwidth is 35°.
3. A 10 GHz antenna with a directivity of 25 dB has conductivity and mismatch losses of 2 dB. Compute the gain of this antenna.
4. Compute the depolarization ratio (in dB) of an antenna with a 75:1 ratio between horizontal and vertical power when measured with a probe.
5. Use a personal computer or calculator to compute data to plot the pattern of a $\lambda_o/2$ dipole antenna.
6. Calculate the VSWR on a 50-ohm coaxial cable that feeds a non-resonant half-wave dipole antenna through a 1:1 balun and a network that tunes out the reactive portion of the antenna's input impedance.

7. Design an 890 MHz Yagi-Uda array antenna for a gain of approximately 10 dB.

8. Design a 1.3 GHz helix antenna for a directivity of 18 dB. Compute the beamwidth, axial ratio and input impedance of the antenna.

9. Design a pyramidal horn antenna for a gain of 20 dB at a frequency of 8 GHz.

10. A parabolic reflector with an F/D ratio of 0.3 is fed by a dipole at the focal point, which is 50 cm from the vertex. Calculate the gain and beamwidth of this antenna at 6 GHz.

11. Design a rectangular microstrip patch antenna for a frequency of 2.4 GHz and calculate the approximate gain and beamwidths. Use a substrate with $\varepsilon_r = 2.0$ and H = 0.318 cm.

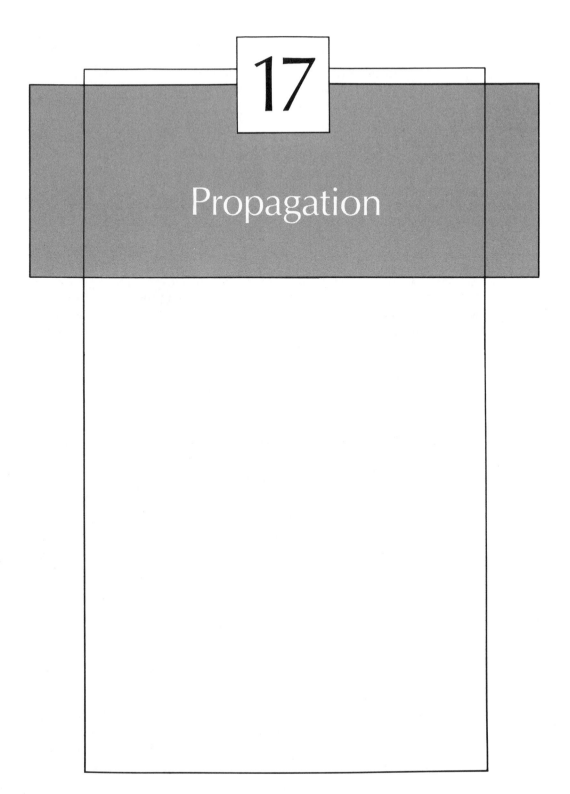

17

Propagation

The propagation medium is an essential part of any communications system. A signal may be sent between two points along a wire, a metallic coaxial cable, a fiber optic cable, or through space. Transmission via wire or metallic cable is governed by RF transmission line theory, while communications over an optical fiber is governed by fiber optic transmission line theory. The subject of this chapter is the transmission of a communications signal through space. Any communications signal transmitted through space requires a transmitting antenna and a receiving antenna. As discussed in Chapter 16, an antenna may be thought of as an impedance matching device between a transmission line and free space. If a signal transmitted through space does not encounter objects or mechanisms that absorb or reflect RF power, then the path from transmitter to receiver (or vice versa) is called free space, and the process is called free space propagation. Reflections from the ground give rise to terrain effects, which may cause fading and/ or multipath transmission. If the signal is high enough in frequency so that atmospheric constituents absorb or reflect RF power, the process is called atmospheric propagation. Tropospheric propagation is the name of the process whereby signals are sent from a transmitter to a receiver via a layer of the atmosphere called the troposphere. If the signal is of a frequency such that it can be reflected off of the ionosphere layer of the atmosphere, the process is called ionospheric propagation. Tropospheric and ionospheric propagation allow communications beyond the optical line of sight, which is limited by the earth's curvature.

17.1 FREE SPACE PROPAGATION

Many communications systems are designed to transfer information directly between two points. Unless tropospheric or ionospheric effects are utilized, such communication is generally limited to line-of-sight in terrestrial systems. Satellite communications is line-of-sight, with the earth station antenna pointed up at the satellite and the satellite antenna pointed toward the earth station. Most line-of-sight systems approximate free space propagation characteristics if the frequencies involved are well into the microwave range.

Free space propagation is governed by the Friis transmission formula, which is given by:

$$\frac{P_r}{P_t} = \frac{\lambda_o^2 G_t G_r}{(4\pi R)^2}$$

where P_r is the received power in watts, P_t is the transmitted power in watts, λ_o is the free-space wavelength in meters, G_t is the gain of the transmitter antenna, G_r is the receiver antenna gain, and R is the distance between receiver and transmitter in meters. The Friis transmission formula assumes that the transmitter and receiver antennas are aligned so that the main beam of each is pointed at the opposite antenna, and also that the antennas are far enough apart so that each is in the far field of the other.

Example 17-1

A 2.2 GHz data link is used to provide communications between two computers in buildings separated by 100 meters. The data link receiver and transmitter have identical antennas with a gain of 15 dB. The transmitter output is 1 watt. Calculate the received power.

Solution

$$\lambda_o = \frac{c}{f} = \frac{3 \times 10^8}{2.2 \times 10^9} = 1.36 \times 10^{-1} \text{ m}$$

$$G_t = G_r = \log^{-1} \frac{G_{dB}}{10} = \log^{-1} \frac{15}{10} = 3.16 \times 10^1$$

$$R = 100 \text{ m}$$

$$P_t = 1.0 \text{ W}$$

$$P_r = \frac{P_t \lambda_o^2 G_t G_r}{(4\pi R)^2} = \frac{(1)(1.36 \times 10^{-1})^2(3.16 \times 10^1)(3.16 \times 10^1)}{(4)(3.14)(100)^2}$$

$$= 1.17 \times 10^{-5} = 11.7 \text{ } \mu W$$

The path loss in a free space transmission system is given by

$$L_p = \left(\frac{4\pi R}{\lambda_o}\right)^2$$

This loss is not a loss in the ohmic sense, but merely describes the spreading of the propagating wave according to the inverse-square law.

Example 17-2

Calculate the path loss in Example 17-1.

Solution

$$R = 100 \text{ m}$$

$$\lambda_o = 1.36 \times 10^{-1} \text{ m}$$

$$L_p = \left(\frac{4\pi R}{\lambda_o}\right)^2 = \left(\frac{(4)(3.14)(100)}{(1.36 \times 10^{-1})}\right)^2 = 8.53 \times 10^7 = 79.3 \text{ dB}$$

Example 17-3

A 6 GHz AM video link is used over a 2 Km path. The gain of the transmitter antenna is 15 dB and the gain of the receiver antenna is 12 dB. If the noise power at the input to the receiver is 50 picowatts and the transmitter power is 5 watts, calculate the signal-to-noise ratio at the receiver input.

Solution

$$\lambda_o = \frac{c}{f} = \frac{3 \times 10^8}{6 \times 10^9} = 5 \times 10^{-2} \text{ m}$$

$$G_t = \log^{-1} \frac{G_{t\text{dB}}}{10} = \log^{-1} \frac{15}{10} = 3.16 \times 10^1$$

$$G_r = \log^{-1} \frac{G_{r\text{dB}}}{10} = \log^{-1} \frac{12}{10} = 1.58 \times 10^1$$

$$R = 2 \times 10^3 \text{ m}$$

$$P_t = 5.0 \text{ W}$$

$$P_r = \frac{P_t \lambda_o^2 G_t G_r}{(4\pi R)^2} = \frac{(5)(5 \times 10^{-2})^2(3.16 \times 10^1)(1.58 \times 10^1)}{[(4)(3.14)(2 \times 10^3)]^2}$$

$$= 9.89 \times 10^{-9} \text{ W}$$

$$\frac{S}{N} = \frac{P_r}{P_n} = \frac{9.89 \times 10^{-9}}{50 \times 10^{-12}} = 1.98 \times 10^2 = 23 \text{ dB}$$

Example 17-4

A standard gain horn with a gain of 15 dB is used as the source antenna on a 30-meter antenna range. The horn is driven by a signal generator with an output of 10 dBm. The antenna under test is a parabolic reflector antenna with a diameter of 1.5 meters. Calculate the received power at the test antenna at a frequency of 3 GHz.

Solution

$$\lambda_o = \frac{c}{f} = \frac{3 \times 10^8}{3 \times 10^9} = 1 \times 10^{-1} \text{ m}$$

$$G_t = \log^{-1} \frac{G_{t\text{dB}}}{10} = \log^{-1} \frac{15}{10} = 3.16 \times 10^1$$

$$G_r = \frac{4\pi\eta_a A_p}{\lambda_o^2} = \frac{4\pi\eta_a\pi \left(\dfrac{D}{2}\right)^2}{\lambda_o^2}$$

Assume $\eta_a = 0.5$

$$G_r = \frac{(4)(3.14)(0.5)(3.14)\ (1.5/2)^2}{(1 \times 10^{-1})^2} = 1.11 \times 10^3$$

$$10 \log \frac{P_t}{1 \times 10^{-3}} = 10 \text{ dBm}$$

$$P_t = 1 \times 10^{-2} \text{ W}$$

$$R = 30 \text{ m}$$

$$P_r = \frac{P_t \lambda_o^2 G_t G_r}{(4\pi R)^2} = \frac{(1 \times 10^{-2})(1 \times 10^{-1})^2(3.16 \times 10^1)(1.11 \times 10^3)}{[(4)(3.14)(30)]^2}$$

$$= 2.47 \times 10^{-5} \text{ watts} = 24.7 \ \mu\text{W}$$

Example 17-5

An 890 MHz FM communications link uses Yagi-Uda array antennas with a gain of 10 dB at both ends of the link. The receiver bandwidth is 15 KHz and the measured noise spectral density at the receiver input is 2.5×10^{-14} volts^2/Hz. The receiver input impedance is 50 ohms. The transmitter output power is 10 watts and the path length is 30 km. Calculate the signal to noise ratio at the receiver input.

Solution

$$\frac{N_o}{2} = 2.5 \times 10^{-14} \text{ volts}^2/\text{Hz}$$

$$N_o = 5.0 \times 10^{-14} \text{ volts}^2/\text{Hz}$$

$$BW = 15 \times 10^3 \text{ Hz}$$

$$Z_i = 50.0 \text{ ohms}$$

$$P_n = \frac{N_o BW}{Z_i} = \frac{(5 \times 10^{-14})(15 \times 10^3)}{50} = 1.5 \times 10^{-11} \text{ watts}$$

$$\lambda_o = \frac{c}{f} = \frac{3 \times 10^8}{890 \times 10^6} = 0.34 \text{ m}$$

$$G_t = G_r = \log^{-1} \frac{10}{10} = 10$$

$$R = 30 \times 10^3 \text{ m}$$

$$P_r = \frac{P_t \lambda_o^2 G_t G_r}{(4\pi R)^2} = \frac{(10)(0.34)^2(10)(10)}{[(4)(3.14)(30 \times 10^3)]^2} = 8.14 \times 10^{-10} \text{ W}$$

$$\frac{S}{N} = \frac{P_r}{P_n} = \frac{8.14 \times 10^{-10}}{1.5 \times 10^{-11}} = 54.3 = 17.4 \text{ dB}$$

17.2 TERRAIN EFFECTS

A communications signal may be subject to loss due to the effect of trees, buildings, or hills. A signal may also be affected by reflections from the ground or water. A reflected signal is often of different phase than the direct signal and therefore may either add to or subtract from it, resulting in fading and/or multipath transmission, which can result in distortion, especially in FM systems.

The effects of obstacles may sometimes be effectively eliminated by placing the receiver and the transmitter antennas at heights that provide adequate terrain clearance. Adequate clearance is usually taken to be the so-called Fresnel clearance given by:

$$h = \left[\frac{\lambda_o}{\dfrac{1}{d_1} + \dfrac{1}{d_2}} \right]^{\frac{1}{2}}$$

In this expression, h is the necessary clearance above the obstacle in meters, d_1 is the distance from the transmitter antenna to the obstacle in meters, d_2 is the distance from the receiver antenna to the obstacle in meters, and λ_o is the free-space wavelength in meters; a flat earth is assumed.

A line-of-sight path that provides clearance equal to or greater than the Fresnel clearance will experience path loss with ± 0.5 dB of the value given by the Friis transmission formula. If the clearance is exactly zero, an additional attenuation of approximately 6 dB will be experienced. If the line-of-sight path is below (i.e., cuts through) the obstacle, the signal will not be totally blocked but will experience considerable attenuation. A line-of-sight path that is below an obstacle by a distance equal to the Fresnel clearance will experience additional attenuation of approximately 17 dB, and a path that is two Fresnel clearances below an obstacle will experience about 23 dB of additional attenuation.

Example 17-6

A 6 GHz data link is to be established between two buildings separated by 5 Km, each 100 meters in height. A third building, 120 meters in height, is in the line-of-sight path of the communications link and is 2 Km from the first building. Calculate the necessary height of the data link transmitter and receiver antennas above the buildings for minimal effect from the third building.

Solution

The transmitter and receiver antennas should be mounted on towers that are one Fresnel clearance distance above the buildings.

$$\lambda_o = \frac{c}{f} = \frac{3 \times 10^8}{6 \times 10^9} = 5 \times 10^{-2} \text{ m}$$

$$d_1 = 2 \text{ Km} = 2 \times 10^3 \text{ m}$$

$$d_2 = 3 \text{ Km} = 3 \times 10^3 \text{ m}$$

$$h = \left[\frac{\lambda_o}{\frac{1}{d_1} + \frac{1}{d_2}} \right]^{\frac{1}{2}} = \left[\frac{5 \times 10^{-2}}{\frac{1}{2 \times 10^3} + \frac{1}{3 \times 10^3}} \right]^{\frac{1}{2}} = 7.75 \text{ m}$$

The line-of-sight path should be 7.75 meters above the obstacle, which is 20 meters above the buildings that house the transmitter and receiver; so towers of 27.75 meters would be required.

Example 17-7

A 1.3 GHz FM communications link is established between two points separated by a distance of 10 Km. A hill, 70 meters in height, lies between the two points. The hill is midway between the two points in the link. Identical transceivers and antennas are used at both points with an output power of 10 watts and an antenna gain of 15 dB. Each of the antennas is mounted on a tower at a height of 20 meters. Estimate the path loss for this link.

Solution

$$\lambda_o = \frac{c}{f} = \frac{3 \times 10^8}{1.3 \times 10^9} = 2.3 \times 10^{-1} \text{ m}$$

$$d_1 = 5 \text{ Km} = 5 \times 10^3 \text{ m}$$

$$d_2 = 5 \text{ Km} = 5 \times 10^3 \text{ m}$$

$$h = \left[\frac{\lambda_o}{\frac{1}{d_1} + \frac{1}{d_2}} \right]^{\frac{1}{2}} = \left[\frac{2.3 \times 10^{-1}}{\frac{1}{5 \times 10^3} + \frac{1}{5 \times 10^3}} \right]^{\frac{1}{2}} = 24 \text{ m}$$

The line-of-sight path is 50 meters below the obstacle (hill) or approximately two Fresnel clearances. The additional attenuation over the free space path loss will be approximately 23 dB.

The free space path loss is

$$L_p = \left[\frac{4\pi R}{\lambda_o}\right]^2 = \left[\frac{(4)(3.14)(10 \times 10^3)}{(2.3 \times 10^{-1})}\right]^2 = 2.98 \times 10^{11} = 114.8 \text{ dB}$$

The total path loss is the sum of the free space loss and the additional loss due to the obstacle or

$$L_T = 114.8 \text{ dB} + 23 \text{ dB} = 137.8 \text{ dB}$$

The concept of Fresnel clearance may also be used to set the height of antennas when no obstacle exists between transmitter and receiver. A transmission path at a distance above level ground at least equal to the Fresnel clearance will minimize facing and multipath effects due to the ground. The obstacle may be assumed to be midway between the antennas in such a case.

Example 17-8

Calculate the antenna height necessary to minimize ground effects in a 150 MHz FM communications link 30 Km long.

Solution

$$\lambda_o = \frac{c}{f} = \frac{3 \times 10^8}{150 \times 10^6} = 2 \text{ m}$$

$$d_1 = d_2 = 15 \text{ Km} = 15 \times 10^3 \text{ m}$$

$$h = \left[\frac{\lambda_o}{\frac{1}{d_1} + \frac{1}{d_2}}\right]^{\frac{1}{2}} = \left[\frac{2}{\frac{1}{5 \times 10^3} + \frac{1}{5 \times 10^3}}\right]^{\frac{1}{2}} = 122 \text{ m}$$

The antennas would need to be mounted on rather tall towers or buildings of suitable height. At this frequency, it might be cost-effective to accept some ground effects and increase power to compensate rather than to mount the antennas at a height of 122 meters.

Terrain effects that result in fading often may be minimized by a technique known as diversity. A communications system may utilize space diversity, frequency diversity, polarization diversity, or some combination of these techniques. Space diversity utilizes two or more transmitting and/or receiving antennas with sufficient physical separation so that if one link experiences a significant fade, the system can automatically switch to the other link. Frequency diversity accomplishes the same task by utilizing two transmitters and receivers at different frequencies. Polarization diversity takes advantage of the fact that a fade may be deeper for one antenna polarization than for another, and automatic switching is arranged to maximize the signal. Diversity systems are more complex and expensive than non-diversity systems, but their installation may often be justified in critical applications.

17.3 ATMOSPHERIC PROPAGATION

The concept of free-space propagation depends on the assumption that the atmosphere is uniform and does not absorb the signal. The concept also disregards terrain effects such as those discussed in Section 17.2. This section examines the effects of atmospheric absorption, and Section 17.4 will discuss the consequences of a non-uniform atmosphere.

The gasses in the atmosphere absorb energy from electromagnetic waves. The only gaseous absorption of importance to communications systems at microwave and lower frequencies is that caused by oxygen and water vapor. Oxygen exhibits its maximum absorption at a wavelength of approximately 0.5 cm or a frequency of 60 GHz. Water vapor exhibits its maximum absorption at a wavelength of approximately 1.3 cm or a frequency of 23 GHz. Since the vast majority of communications systems operate at frequencies less than 18 GHz, these absorption maxima are of little concern. Of more interest to communications system designers is the attenuation due to atmospheric absorption at frequencies below 18 GHz. For typical conditions (20°C, 70% relative humidity), the total atmospheric absorption due to oxygen and water vapor is approximately 0.02 dB/Km over a frequency range of 12 GHz to 18 GHz, approximately 0.015 dB/Km from 4 GHz to 12 GHz, and less than 0.01 dB/Km below 4 GHz.

Example 17-9

An 8-GHz data link is used to provide digital voice communications between two government facilities located 40 Km apart. Calculate the loss due to atmospheric absorption over this link.

Solution

$$L = 0.015 \text{ dB/Km} \times 40 \text{ Km} = 0.6 \text{ dB}$$

Thus atmospheric absorption introduces an additional 0.6 dB loss over and above the free-space path loss.

Atmospheric attenuation caused by rain or fog is normally much more important than gaseous absorption at frequencies above 4 GHz. Atmospheric attenuation due to rain or fog is not primarily due to absorption, but is the result of scattering by the water droplets. Heavy rain (16 mm/hr) results in approximately 0.3 dB/Km attenuation at 10 GHz, while a light rain (2 mm/hr) or light fog (visibility 100 m) results in approximately 0.015 dB/Km attentuation at 10 GHz. Below 4 GHz, the effects of rain and fog are less than 0.01 dB/Km and are usually neglected.

Example 17-10

A 10 GHz FM communications link operates over a path length of 30 Km. Calculate the loss over and above the free-space path loss when the system is operated during a heavy rain storm occupying the entire path length.

Solution

The atmospheric absorption loss is 0.015 dB/Km; the attenuation due to rain is 0.3 dB/Km. The total attenuation is 0.315 dB/Km or

$$L = 0.315 \text{ dB/Km} \times 30 \text{ Km} = 9.5 \text{ dB}$$

It should be noted that areas of heavy rainfall are normally localized and would not extend over the entire 30 Km path length; thus the attenuation could be considerably less than 9.5 dB.

In telecommunications systems where repeaters are used, the spacing specified will depend strongly upon the rainfall statistics for the area and the allowable down time for the system. At higher frequencies ($\approx$18 GHz) repeater spacing may be limited to a few kilometers.

Diversity, especially space diversity, is often used to combat the effects of heavy rainfall.

17.4 TROPOSPHERIC PROPAGATION

The earth's atmosphere is normally divided into four regions: the troposphere, which extends from the ground to a height of 10 to 15 Km; the tropopause, a transition zone ranging from approximately 5 Km at the two poles to approximately 20 Km at the equator; the stratosphere, which extends to approximately 50 Km above the surface; and finally the ionosphere. In the troposphere, the temperature generally decreases with altitude, while in the stratosphere, it is relatively constant. The troposphere contains most of the clouds responsible for active weather. In the troposphere, the index of refraction (a measure of the ability of a medium to bend electromagnetic waves) decreases with height.

If the index of refraction decreases with height in a linear fashion, electromagnetic waves are bent downward, so that they can be received beyond the optical line of sight. This effect is equivalent to assuming that the earth's radius is increased or that the earth is effectively flattened. Under normal atmospheric conditions, the earth's effective radius is increased by a factor of approximately 1.3. Under special conditions that cause the index of refraction to decrease rapidly, the effective radius of the earth can become infinite, which implies a flat earth for electromagnetic propagation. Variations in index of refraction with height are a strong function of meteorological conditions such as temperature and humidity, so that successful communication via tropospheric bending is difficult to predict.

A second tropospheric propagation mode, called tropospheric scatter, or troposcatter, is more predictable and is widely used. In troposcatter, electromagnetic energy is scattered in a forward direction by a volume in the troposphere, where the transmitter and receiver antenna beams intersect. The scattering is the result of irregular variations in the index of refraction of the atmosphere. Troposcatter communications may extend to distances over 1000 Km. This mode has been utilized in systems operating from 50 MHz to 10 GHz. Successful systems are often high power (50 KW) and utilize high gain (40 dB) transmitting and receiving antennas at low angles (3°). A troposcatter signal is subject to fading and therefore space, frequency, and/or polarization diversity are often useful in improving the signal levels existing on the link. A propagation mechanism in some ways similar to troposcatter is meteor burst propagation in which signals propagate as a result of scattering from meteorites.

17.5 IONOSPHERIC PROPAGATION

Above the stratosphere is the ionosphere, ranging from approximately 50 Km to 300 Km in height. The ionosphere consists of several layers of charged particles or ions, which can reflect (sharply bend) electromagnetic energy (generally below 30 MHz) back down to the earth's surface to greatly increase the communications range over that attainable by line-of-sight systems. The characteristics of the ionosphere are primarily influenced by ultraviolet light and by the charged particles emitted by the sun. The lowest useful layer of the ionosphere is called the E layer, whose average height of maximum ionization is about 110 Km. A layer below the E layer, called the D layer, absorbs electromagnetic energy below approximately 5 MHz and thus prevents extended communications range when it is present. The D and E layers only exist when sunlight is present. The ionospheric layer most useful for long distance communication is the F layer, which splits into F_1 and F_2 layers during the day and merges into a single layer at sunset. The F_1 layer has its maximum ionization at a height of approximately 225 Km and the F_2 layer has its maximum at a height of about 320 Km. In the evening, the F layer is about 280 Km above the earth's surface. Because the atmosphere is so thin at the height of the F layer, the ionization, which occurs during the day due to sunlight, decays very slowly during the dark hours. As the ionization decreases, the maximum usable frequency (MUF) for a given path decreases throughout the night.

The maximum distance for a single-hop E layer communications link is approximately 2000 Km. The maximum F layer single-hop link is approximately 4000 Km. Multiple hops, in which the signal is reflected from the ionosphere to the earth's surface and back to the ionosphere before reaching the receiving antenna, may extend the maximum range even further, although the received signal levels will be very low.

Ionospheric propagation is characterized by a limited predictability of maximum usable frequency range and total path loss. In addition, communication via the ionosphere is often subject to deep fades. During periods when the solar flux levels are low, the MUF may be less than 15 MHz. Ionospheric propagation is generally most useful for broadcast, amateur and other non-critical communications systems. In critical systems (such as those used in military applications or data communications) satellite links or troposcatter are used for long distance communications.

EXERCISES

1. A 3 GHz digital radio system is used to provide short range communications for military activities. The transmitter antenna has a gain of 18 dB, and the receiver antenna has a gain of 12 dB. If the transmitter and receiver are separated by 1 Km and the transmitter output power is 5 watts, calculate the received power.
2. Calculate the path loss in dB for Exercise 1.
3. A 2 GHz digital video link is used over a 5 Km path. The transmitter has a power of 10 watts and an antenna gain of 17 dB. The receiver has an input noise power of 50 picowatts and an antenna gain of 20 dB. Calculate the signal-to-noise ratio at the receiver input.
4. The digital video link of Exercise 3 has a 200-meter hill halfway between the receiver and transmitter. Calculate the necessary height for the transmitter and receiver antennas.
5. Calculate the antenna height necessary to minimize ground effects in an 890 MHz communications link that is 10 Km long.
6. A 10 GHz digital radio system operates over a path length of 50 Km. Calculate the total path loss for this system if the system is operated during a heavy rainstorm that occupies the entire path length.

18

Fiber Optic Communications Systems

In recent years, fiber optic communications systems have gained an important position in communications technology. A fiber optic system (the most widely used type of optical communications system) consists of an optical transmitter (light source), an optical transmission line (optical fiber) and an optical receiver (photo detector).

The beginnings of optical communications may be traced back to the development of the laser in the early 1960s. Compared to conventional optical sources, laser radiation is essentially monochromatic (single-frequency), coherent (in-phase and unidirectional), and intense. Laser radiation is similar in many ways to microwave radiation, but at frequencies in the terahertz (thousands of GHz) region. Early optical communications experiments utilized unguided transmission through the atmosphere. On a clear day, atmospheric attenuation at optical frequencies is only a few dB per Km, but of course, this attenuation is greatly influenced by conditions such as fog, smoke, clouds, rain and snow. Because of the problems encountered with unguided transmission through the atmosphere, a number of researchers conducted experiments in which a laser beam was confined to a transmission channel with lenses spaced at 10- to 100-meter intervals along the channel. Since such a system was complex, researchers next investigated glass fibers as a possible transmission medium. The early glass fibers, however, exhibited attenuations of the order of 1000 dB/Km and thus were not practical. In the early 1970s, Corning Glass introduced a glass fiber cable with an attenuation of approximately 20 dB/Km that finally made fiber optic communications practical.

Fiber optic systems are used widely in telecommunications, instrumentation, cable television, and data transmission and distribution. These systems feature low loss, high bandwidth, immunity to electromagnetic interference, small size, light weight, non-electrical transmission, and relative security against unauthorized tapping. In addition to these advantages, fiber optic systems are frequently more cost-effective than alternative approaches to communications. Most fiber optic systems are digital in nature, but a few utilize analog modulation techniques when specifications favor this approach. Current fiber optic systems utilize semiconductor sources and detectors and low-loss cables with attenuation levels of a few dB/Km or less.

18.1 OPTICAL CABLES

Optical fibers are fabricated from dielectric materials—in most cases glass (SiO_2) mixed with various dopants to control the optical and mechanical properties of the glass fiber. Optical properties are governed primarily by the index of refraction n of the material, which is related to the relative dielectric constant, ε_r, (for non-magnetic materials) by

$$n = \sqrt{\varepsilon_r}$$

Example 18-1

A glass fiber optic cable has a relative dielectric constant of 2.10. Calculate the index of refraction for this cable.

Solution

$$n = \sqrt{\varepsilon_r} = \sqrt{2.10} = 1.45$$

To understand how a transparent dielectric medium can function as an optical transmission line, refer to Snell's law, which describes refraction of a ray of light passing from a medium of higher refractive index (n_1) into a medium of lower refractive index (n_2). Figure 18-1 illustrates this situation. Snell's law states:

$$n_1 \sin \theta_1 = n_2 \sin \theta_2$$

In the above expression, n_1 is the index of refraction in medium 1, θ_1 is the incidence angle, n_2 is the index of refraction in medium 2, and θ_2 is the angle of refraction. As the ratio n_1 to n_2 increases (for a given θ_1), a point will be reached where $\theta_2 = 90°$. As this ratio is increased further, a phenomenon known as total internal reflection will occur. For fixed n_1 and n_2, the incidence angle θ_1 may be varied until $\theta_2 = 90°$. This angle is called the critical angle, θ_c. for $\theta_1 > \theta_c$, total internal reflection again occurs.

Example 18-2

A ray of light in a medium of relative dielectric constant $\varepsilon_r = 2.2$ passes into a second medium with relative dielectric constant $\varepsilon_r = 2.1$ at an incidence angle of 45°. Calculate the angle of refraction in the second medium.

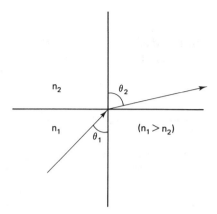

Figure 18-1   Refraction at a dielectric interface with $n_1 > n_2$.

Solution

$$n_1 = \sqrt{\varepsilon_{r_1}} = \sqrt{2.2} = 1.48$$

$$\theta_1 = 45°$$

$$n_2 = \sqrt{\varepsilon_{r_2}} = \sqrt{2.1} = 1.45$$

$$n_1 \sin \theta_1 = n_2 \sin \theta_2$$

$$\sin \theta_2 = \frac{n_1}{n_2} \sin \theta_1 = \frac{1.48}{1.45} (0.707) = 0.722$$

$$\theta_2 = \sin^{-1}(0.722) = 46.2°$$

In the case of an optical fiber, an inner core of refractive index n_1 is covered by an outer cladding of refractive index n_2, where $n_1 > n_2$. The end face of the fiber is cut at right angles to the fiber axis. A ray of light entering the end of the fiber will propagate down the fiber through a series of total internal reflections, provided θ_1 is greater than the critical angle θ_c. This propagation mode for a single ray of light is illustrated in Figure 18-2. Because the range of angles over which total internal reflection takes place is limited, the concept of an acceptance cone, defined by an acceptance angle θ_A is useful. The sine of the acceptance angle is called the numerical aperture. The numerical aperture is one of the most important properties of an optical fiber, for it describes the efficiency at which optical signals may be coupled into the fiber. Figure 18-3 illustrates the acceptance cone and associated acceptance angle. An expression for the numerical aperture is:

$$NA = \sqrt{n_1^2 - n_2^2}$$

Example 18-3

Calculate the numerical aperture and acceptance angle for a fiber cable with a core index of refraction of 1.45 and a cladding index of refraction of 1.30.

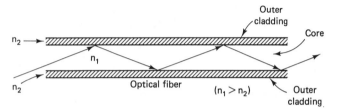

Figure 18-2 Total internal reflection in an optical fiber.

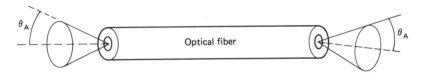

Figure 18-3 Acceptance cone and acceptance angle in an optical fiber.

Solution

$$n_1 = 1.45 \qquad n_2 = 1.30$$

$$NA = \sqrt{n_1^2 - n_2^2} = \sqrt{(1.45)^2 - (1.30)^2} = 0.643$$

$$\theta_A = \sin^{-1}(NA) = \sin^{-1}(0.643) = 40°$$

Light rays propagate at various angles down the fiber as determined by the acceptance angle, θ_A. Rays propagating at small angles are called low-order modes; rays propagating at higher angles are called high-order modes. For a given length of fiber, higher order modes travel a longer path length than low order modes. Optical losses from scattering and absorption are a function of path length, so higher order modes exhibit more loss than low order modes. In addition, the higher order modes require additional time to traverse the longer path, so a low-order ray will arrive at the cable end sooner than a high-order ray—giving rise to a phenomenon known as modal or multipath dispersion. Material dispersion is a function of the variation of the index of refraction with frequency. Both types of dispersion limit the rise time of individual pulses in digital systems and thus the maximum data rate. In analog systems, dispersion limits the bandwidth. Since dispersion is a function of cable length, cables are often characterized by a rise time per unit length (ns/Km) or by a bandwidth-length product (MHz·Km).

In general, cables with high numerical apertures are desirable when connecting two fiber cables together by a splice or with connectors, but dispersion and attenuation generally increase for higher numerical apertures. The loss in dB resulting from coupling two cables with unequal numerical apertures is $\alpha_{NA} = 20$ log (NA_1/NA_2), where α_{NA} is the loss in dB caused by the unequal numerical apertures and NA_1 and NA_2 are the numerical apertures of cable 1 and cable 2 respectively.

Example 18-4

A fiber optic cable with a numerical aperture of 0.65 is to be spliced to a second fiber optic cable with a numerical aperture of 0.42. Calculate the coupling loss in dB resulting from the unequal numerical apertures.

Solution

$$NA_1 = 0.65 \qquad NA_2 = 0.42$$

$$\alpha_{NA} = 20 \log \frac{NA_1}{NA_2} = 20 \log \frac{0.65}{0.42} = 3.79 \text{ dB}$$

The loss in dB resulting from coupling two cables with unequal core diameters is $\alpha_A = 20 \log (D_1/D_2)$, where α_A is the area loss in dB due to the unequal core diameters and D_1 and D_2 are the diameters of core 1 and core 2 respectively.

Example 18-5

A fiber optic cable with a core diameter of 100 μm is to be spliced to a cable with a core diameter of 80 μm. Calculate the coupling loss from the unequal diameters.

Solution

$$D_1 = 100 \text{ μm} \qquad D_2 = 80 \text{μm}$$

$$\alpha_A = 20 \log \frac{D_1}{D_2} = 20 \log \frac{100}{80} = 1.94 \text{ dB}$$

The propagation delay of a fiber cable is given by $t_p = n_1 l/c$, where t_p is the propagation delay in seconds, n_1 is the core index of refraction, l is the length of the cable in meters, and c is the speed of light in meters per second.

Example 18-6

Calculate the propagation delay for a 1 Km fiber optic cable with a core index of refraction of 1.455.

Solution

$$t_p = \frac{n_1 l}{c} = \frac{(1.455)(1 \times 10^3)}{(3 \times 10^8)} = 4.9 \times 10^{-6} = 4.9 \text{ μsec}$$

This can be compared to a free-space propagation delay of 3.3 μsec.

Three types of fiber optic cables are in use: the step index cable, the graded index cable, and the single mode cable. The step index cable is constructed with an inner core of one index of refraction and an outer cladding with a lower index of refraction as discussed earlier in this section. Step index cables support multiple modes and are characterized by large core diameters, easy coupling to light sources, and low-cost connectors.   The graded index fiber optic cable exhibits a continuous variation in index of refraction as a function of radial distance.   A graded index cable is more difficult to manufacture than a step index cable and is quite expensive, but it exhibits less dispersion and thus is capable of supporting a higher data rate (or bandwidth) than a step index cable.   The state-of-the-art in fiber optic cables is the single mode or mono-mode cable.   Single mode cables have a small diameter in comparison to the light wavelength, so that only a single mode is supported. Single mode cables are low-loss, low-dispersion and durable (greater bending radius than other cable types), but are relatively expensive.   Single mode cables are most cost-effective on long-haul telecommunications links.   All fiber optic cable types are available with outer protective jackets.

Three wavelengths are used in fiber optic systems, 0.85 μm, 1.3 μm, and 1.55 μm.   The 0.85 μm wavelength is the most popular, since low-cost sources are available at this wavelength.   The 1.3 μm wavelength results in minimum dispersion in optical cables and is used for high data rate or wide bandwidth applications. The 1.55 μm wavelength results in minimum attenuation, so it is sometimes used on long-haul telecommunications links.   The popular 0.85 μm multi-mode step index fiber optic cable typically exhibits an attenuation of approximately 2 dB/Km and a bandwidth of approximately 40 MHz·Km.   A 0.85 μm graded index cable exhibits about the same attenuation, but its bandwidth increases to approximately 500 MHz·Km.   At 1.3 μm, a multi-mode cable exhibits an attenuation of approximately 1 dB/Km with a bandwidth of over 5 GHz·Km.   A single mode cable at this wavelength has approximately 0.4 dB/Km loss.   At 1.55 μm, a typical single mode fiber has less than 0.2 dB/Km attenuation and a bandwidth of over 5 GHz·Km. The bandwidth of some high quality single mode fibers is over 100 GHz·Km.

18.2 FIBER OPTIC CABLE CONNECTORS AND SPLICES

A practical but extremely important consideration in any fiber optic system is the need for connectors and splices.   The extremely small cross sections of the fibers require precise alignment.   Coupling loss increases rapidly as displacement between the two ends of the fibers increases.   In addition, additional loss is incurred if the ends are rough and not parallel.   The coupling problem is most challenging for single mode fibers because of their small diameters.

Connectors for fiber optic cables are available from several manufacturers. The most popular type of fiber optic connector is similar in style and size to the popular microwave SMA connector.   Connectors are designed to minimize lateral and angular displacement of the two fibers.   Typical connector loss is 0.5 dB. Special tool kits are available to assist in connector installation.

Fiber optic cable splices are typically executed by means of an electrical arc or by thermal fusion.   Special equipment is available from manufacturers to execute splices with losses of 0.3 dB or less. Couplers, created by splicing several fiber cables together, can be fabricated with thermal fusion equipment.

18.3 OPTICAL SOURCES

An ideal light source for a fiber optic communications system would exhibit a long lifetime, high efficiency, low cost, adequate power output and in addition be easily modulatable. The two types of sources in common use are LED (light emitting diode) sources and laser sources. The LED, which is an incoherent light source, is characterized by low power output (≈ -15 dBm with coupling loss) and low modulation bandwidth (≈ 50 MHz), but is relatively low in cost and exhibits a long lifetime. A laser light source is monochromatic and coherent, which makes it ideal for single mode fiber applications. Gas lasers are generally not used in fiber optic systems because of their bulk, expense, and low efficiency. Semiconductor laser diodes, however, find wide application in fiber optic systems, since they exhibit high power output ($\approx +5$ dBm with coupling loss) and a high modulation bandwidth (≈ 1 GHz). The disadvantages of the laser diode are high cost, relatively short lifetimes, and the need for temperature and power stabilization circuitry. Single mode fiber optic cables start to exhibit non-linearity at power levels of approximately $+10$ dBm and multimode fibers can be driven with $+20$ dBm or more before non-linearity is evident, so the above two sources allow linear operation within all cable types. The GaAs LED emits light at approximately 0.85 μm and is widely used as source at this wavelength. At the longer wavelengths of 1.3 μm and 1.55 μm, LEDs fabricated from semiconductor materials such as InGaAs are used as sources. Laser diodes made from a wide variety of semiconductor materials are also popular at these wavelengths.

Semiconductor sources may be easily modulated by varying the bias current to the device in the case of analog modulation techniques or simply switching the bias current on and off in the case of digital modulation. More complex modulation schemes are also in use.

18.4 OPTICAL DETECTORS

The purpose of the optical detector in a fiber optic communications system is to convert the optical signal received from the fiber into an electrical signal. At 0.85 μm, Si PIN photodiodes, Ge avalanche photodiodes and GaAs MESFET transistors are used as optical detectors. At 1.3 μm and 1.55 μm, the only optical detector in widespread use is the Ge avalanche photodiode.

The Si PIN photodiode is simple, low in cost, and reliable, but may only be used in 0.85 μm systems. The Ge avalanche photodiode offers improved performance over the Si PIN photodiode, but a system utilizing this device is more costly, since high bias voltages are required. In addition, Ge avalanche photodiodes are inherently non-linear devices so the only modulation technique applicable is digital modulation.

The GaAs MESFET transistor may be used as an optical detector by optically coupling the fiber cable to the semiconductor chip surface. A GaAs MESFET detector is a very low noise device and has the desirable property of exhibiting gain. The effort required to optically couple the fiber to the device is considerable, however, and manufacturing cost is high.

18.5 FIBER OPTIC SYSTEMS

A typical fiber optic communications system consists of a digitally modulated source, a fiber optic cable with connectors and a detector. Example 18-7 illustrates the link calculations for such a system.

Example 18-7

A Texas Instruments TIXL472 LED is the source in a 2 Km fiber optic communications link. The fiber optic cable is available in 0.5 Km rolls and exhibits a loss of 2 dB/Km at the LED emission wavelength of 0.91 μm. The receiver noise floor is -40 dBm. Calculate the signal-to-noise ratio at the receiver.

Solution

The data sheet for the TIXL472 indicates that its output power is 1 mw (0 dBm) at a bias current of 50 ma. This optical source is terminated in an optical connector with an LED to fiber coupling loss of 20 dB.

Assume the two required connectors introduce 0.5 dB loss each and that each splice introduces 0.3 dB loss. The link loss is:

LED to fiber coupling	20.0 dB
Source connector	0.5 dB
Fiber length	4.0 dB
3 splices	0.9 dB
Detector connector	0.5 dB
Fiber to detector coupling	0.2 dB
TOTAL	26.1 dB

The power available at the receiver (actually signal + noise power, but can be approximated as signal power alone) is the LED power less the link loss:

$$0 \text{ dBm} - 26.1 \text{ dB} = -26.1 \text{ dBm}$$

This power level is 13.9 dB above the noise floor of the receiver, so the signal-to-noise ratio is approximately 13.9 dB.

EXERCISES

1. Calculate the index of refraction for a glass fiber optic cable with a relative dielectric constant of 2.3.

2. A glass fiber optic cable has an index of refraction of 1.6. Calculate its relative dielectric constant.

3. A ray of light in a medium of relative dielectric constant $\varepsilon_r = 2.3$ passes into a second medium with relative dielectric constant 1.9 at an incidence angle of 30°. Calculate the angle of refraction in the second medium.

4. Calculate the numerical aperture for a fiber cable with a core index of refraction of 1.6 and a cladding index of refraction of 1.4.

5. Calculate the acceptance angle for problem 6.

6. A fiber cable has an acceptance angle of 30° and a core index of refraction of 1.4. Calculate the cladding index of refraction.

7. A fiber optic cable with a numerical aperture of 0.8 is to be spliced with a cable with a numerical aperture of 0.6. Calculate the coupling loss in dB.

8. A fiber optic cable with an acceptance angle of 40° is to be spliced with a cable with an acceptance angle of 50°. Calculate the coupling loss in dB.

9. A fiber optic cable with a core diameter of 120 μm is to be spliced with a cable with a core diameter of 80 μm. Calculate the coupling loss in dB.

10. An 0.5 Km fiber optic cable has a propagation delay of 3.0 μsec; calculate its core index of refraction.

11. A TIXL472 LED is used in a 10 Km fiber optic communications link. The fiber optic cable is available in 500m rolls and exhibits a loss of 1.6 dB/Km at 0.91 μm. The receiver noise floor is −60 dBm. Calculate the signal-to-noise ratio at the receiver.

19

Survey of Communications Systems

To this point, this text has emphasized components and subsystems of communications systems. This chapter provides a brief introduction to and survey of some practical systems. It is intentionally general in nature; communication system technology is constantly and rapidly changing, and the interested reader should consult current systems manuals for a detailed knowledge of a specific transmitter, receiver, or communications system.

The chapter will provide a discussion of various terrestrial communications systems, satellite communications systems, and digital data communications systems. The organization of the topics is somewhat arbitrary and some overlap exists. For example, data communications are carried by both terrestrial and satellite links, and fiber optic systems can carry both analog signals and digital data.

19.1 BROADCAST COMMUNICATIONS SYSTEMS

The terrestrial communications systems that are probably the most familiar to readers of this text are the AM, FM and TV broadcast systems.

In the United States and most other countries, the most frequently used frequency band for AM broadcasting is the 540 KHz to 1600 KHz range. Portions of the HF frequencies from approximately 2 MHz to 20 MHz are also used worldwide for international AM broadcasting. AM broadcast transmitters are characterized by amplitude modulation, output powers of the order of several thousand watts and quarter-wavelength tall towers for the antennas. High-power AM broadcast transmitters generally utilize quartz crystal-controlled solid state oscillators and solid state low-level stages followed by high power vacuum tube amplifiers in their final stages to generate the required power. AM broadcast receivers are characterized by simplicity and low cost. AM broadcasting is restricted to narrow bandwidths to conserve the spectrum available for this service and thus is relatively low fidelity.

The FM broadcast band, from 88 MHz to 108 MHz, is wide enough so that increased transmitted signal bandwidth is possible. This increased bandwidth permits the utilization of the inherent immunity to noise of wide-band frequency modulation and a high fidelity broadcast service thus becomes possible. FM broadcast transmitters are similar to AM transmitters in that the oscillator and low-level stages are usually solid state, while the output stages utilize vacuum tubes. Most FM broadcast transmitters are designed for several thousand watts of output power. Some low-power FM transmitters are available with a 100% solid state design. FM transmitting antennas are generally mounted atop large towers. FM receivers span the range from very low-cost, simple portable designs to extremely complex, expensive high fidelity designs.

The frequency bands used for television broadcasting include a pair of VHF bands (channels 2–13) and a group of UHF bands (channels 14–88). A broadcast television signal differs from a simple AM or FM signal in that it must contain both audio and video information.   The audio information in a broadcast television signal utilizes FM modulation techniques, while the video portion utilizes a modified AM technique known as vestigial sideband modulation.   TV transmitters generally have outputs of several thousand watts, and their antennas are similar to FM transmitter antennas.   Television transmitter and receiver systems are quite complex; the interested reader should consult a textbook specifically devoted to that subject for further information.

The technology utilized in AM, FM, and TV broadcast systems is, for the most part, a very mature technology.   Advances in integrated circuits periodically allow improved designs, but only a very small group of designers is involved in this effort.

19.2 CABLE TELEVISION SYSTEMS

Cable television systems have made a major impact upon the broadcast television industry in recent years.   Cable systems offer consumers the choice of dozens of channels with special interest programming.   In a cable TV system, signals received off the air or via a satellite earth station are frequency-division multiplexed onto one or two coaxial cables and distributed to customers.   The "channels" utilized by cable TV systems are in some cases the same frequencies as for broadcast TV, but in other cases, frequencies allocated to non-broadcast services are also used. All signals are confined to the coaxial cable (if the system is properly designed and maintained) and do not cause interference to these services.   Premium channels are generally "scrambled" prior to transmission and require a decoder at the customer's receiver.   Cable TV systems utilize relatively mature technology at present, but have the potential for some rather exciting data communications applications in the future.

19.3 MOBILE COMMUNICATIONS SYSTEMS

It has been frequently said that we are a mobile society.   We are also a society that requires instantaneous communication, so it is not surprising that mobile communications is one of the most important current applications of communications technology.   Millions of such systems are installed in automobiles, trucks, boats, and aircraft.   Users of mobile systems include business, the military, public service organizations (fire and police, for example) and the general public.

Most mobile communications occurs between 100 MHz and 1 GHz, although a few services, such as the U.S. citizen's band, utilize frequencies as low as 26 MHz.   Most mobile systems use FM modulation because of its inherent noise immunity, but a few AM and SSB systems continue to exist.   Amplitude compandered single sideband (ACSSB) modulation is also finding favor in mobile communications.   Power outputs from mobile transceivers (transmitters-receivers) typically range from 5 watts to 50 watts. A major problem in mobile communications is fading. Mobile communications systems, especially those mounted in automobiles or trucks and operated in a large city, encounter numerous obstacles that cause deep fades of 40 dB or more. These fades, when combined with relatively low transceiver power outputs, severely limit communications range between a base station (normally having higher power output and a much more favorable antenna location) and a mobile system. Direct communication between two mobile systems is nearly impossible unless the vehicles are very close to each other.

One common solution to the range problem is to establish a repeater system with antennas mounted as high as practical. The repeater consists of a receiver/antenna, which picks up the mobile signal, and a separate transmitter/antenna, which retransmits the signal to other mobile and/or base stations. The repeater's receiver and transmitter must operate at different frequencies to minimize interference (the receiver and transmitter operate simultaneously). Repeaters must have receiving and transmitting antennas physically separated by a considerable distance or must have a rather involved band pass filtering scheme at the receiver input to avoid desensitization ("de-sense") by the transmitter. Repeater transmitter power output levels may be several hundred watts. Marine communication is normally not subject to major obstacles and thus repeaters are often unnecessary. Aircraft communications systems have the advantage of an obstacle-free line-of-sight path and an extremely high antenna, so long distance communication (several hundred kilometers) is possible without the use of repeaters. It is interesting to note that mobile radio services utilizing geosynchronous satellites as repeaters are rapidly evolving.

19.4 CELLULAR TELEPHONE SYSTEMS

A cellular telephone is a mobile communications system connected to the public telephone network. To the user, a cellular telephone system appears very similar to a telephone wired to the public telephone network. Cellular telephone systems divide an urban area into hundreds of "cells." At a central location within each of these cells, a very low power repeater is established. Because of the low powers involved and the availability of hundreds of frequency pairs, inter-repeater interference is not a problem in these systems.

The mobile units and repeaters are under the control of a centralized digital computer that automatically switches the mobile unit from cell to cell as it moves through the urban area. The centralized computer also controls the connection to the public telephone network and maintains accounting information. The user of the cellular system is unaware of the complex switching required to implement this mobile telephone system. Hand-held cellular telephones are available for individuals who have a need for communications while on foot.

19.5 PAGING SYSTEMS

The expense and bulk of a cellular telephone cannot often be justified, yet certain individuals require some means of notification that they should contact someone else via telephone. This need is met by pocket paging systems. In this type of communications system, the paging firm typically establishes a high-power transmitter and well-situated antenna at a central location within an urban area. Each subscriber to the service carries a small receiver that has a "squelch" system, which opens only when its unique code is received. Upon receipt of this code, the receiver will "beep" at the subscriber who then locates a public telephone to call in for his message. Some more sophisticated pagers contain a digital display that shows the user a telephone number to call or may contain a short message. Paging systems are normally one-way digital communications systems, although a few are capable of transmitting a voice message from an operator.

19.6 PACKET RADIO SYSTEMS

Most digital communications occurs via the public telephone network, satellites, or dedicated wire or fiber optic lines. In some instances, however, none of these methods is suitable and a microwave/RF communications link is necessary. All data communications systems require some protocol or set of rules so that the receiver can interpret what the transmitter has sent. The packet-switching protocols used in the public data networks are sometimes used in a modified form in radio systems designed for data communications. Packet communications will be discussed later in this chapter. It should be noted that packet radio is not limited to data communications but may also be used for digitized audio, video, or other analog signals.

19.7 LINE-OF-SIGHT SYSTEMS

While many communications systems are line-of-sight systems, the term is usually reserved for the microwave communications links utilized by the public telephone network and by transportation and pipeline companies. Line-of-sight systems are often more cost-effective than wire, coaxial or fiber optic cable links.

Most line-of-sight systems operate in the range of 1 GHz to 10 GHz with a majority operating at approximately 4 GHz. Line-of-sight systems may span several thousand kilometers with repeaters spaced every 50 Km to 100 Km. Repeaters have offset input and output frequencies and are characterized by high-gain directional antennas mounted on high towers and transmitter powers of only a few watts. Although some systems utilize SSB modulation, the majority are FM systems. Frequency division or digital time division multiplexing is often used to combine many telephone and/or data channels onto a single carrier. Because line-of-sight systems are subject to fading, frequency or space diversity is sometimes used to improve reliability. Line-of-sight system repeaters are typically of the nonde-modulation type, which means that the received signal is not translated to baseband and used to modulate the transmitter; the received signal is instead directly translated to the new transmitter frequency. By avoiding the translation to baseband, these repeaters minimize distortion and noise.

19.8 TROPOSPHERIC SCATTER COMMUNICATIONS SYSTEMS

Tropospheric scatter propagation was discussed in Chapter 12. Tropospheric scatter allows reliable communications at distances over 1000 Km. This type of communications has been used most often by the military; it does not depend upon unreliable ionospheric propagation and does not utilize vulnerable repeaters or satellites. Tropospheric scatter communications systems are characterized by extremely high antenna gains and high-power transmitters of 1 KW to 50 KW output power. Frequencies from 100 MHz to 15 GHz have been used for troposcatter, although the range of 100 MHz to 5 GHz has proven most effective for long distance links. Since troposcatter systems are subject to fading, various diversity techniques (space, frequency, or polarization) are frequently used to improve reliability.

19.9 SATELLITE COMMUNICATIONS SYSTEMS

In 1945, science fiction writer Arthur C. Clarke advanced the idea of orbiting satellites providing global communications. Clarke predicted that a satellite placed at an orbital height above the equator such that its rotational period matched that of the earth would appear motionless, and thus could be used for communications without complicated tracking devices. Such an orbit is called a geosynchronous orbit.

Clarke's dream was realized in 1963 with the launching of Syncom, the first geosynchronous satellite. Syncom was followed in the mid-1960s by a number of commercial satellites such as Intelsat I (Early Bird), Anik and Westar. The military also recognized the importance of communications satellites and was actively involved in launches during this period.

It is not always advantageous, however, to place satellites in a geosynchronous orbit. Certain communications satellites as well as weather satellites and remote sensing satellites utilize elliptical or circular orbits to obtain coverage of portions of the earth not possible with geosynchronous orbits.

Early communications satellites used analog technology and frequency division multiplexing techniques. As digital communication became more important in the late 1970s and early 1980s, analog satellite channels were used for digital transmission. The first commercial satellite designed specifically for digital data transmission was the Satellite Business Systems (SBS) satellite.

Nearly all of the communications principles discussed in earlier chapters of this text may be applied to satellite communications. The communications satellite may be viewed as a microwave repeater in the sky.

19.10 THE PUBLIC TELEPHONE NETWORK

The most pervasive communications system in existence today is the public telephone network. In most developed countries, a significant percentage of the population has convenient access to this system. Although the system was originally designed for analog communications, it is increasingly being used for digital data communications and is utilizing digital transmission techniques even for sending analog information.

Because of the huge investment in equipment and lines, the public telephone network in various countries has tended to lag behind current technology and has evolved slowly. The most dramatic recent change in the network has been an evolution from outdated electromechanical switches, to the use of reed relays, to the use of digital electronic switching using PCM techniques.

Internationally accepted specifications for telephone networks include a frequency response of 300 Hz to 3400 Hz, harmonic distortion down at least 26 dB and a signal-to-noise ratio of at least 30 dB. These specifications differ greatly from those of a high-quality audio system and result from technical and financial trade-offs. Since the network is also used for data communications, an additional specification, group delay, is required. Group delay is the rate of change of phase shift with frequency. This specification is of little importance for voice signals but is critical for data communications.

The public telephone network generally consists of three types of links, depending upon the distance involved and the number of communications channels required. The local connection or route is the baseband link that joins an individual subscriber to a so-called local exchange and is often referred to as the "local loop." A toll-connecting trunk joins the local exchanges to a centralized switching center. An intertoll trunk connects switching centers both nationally and internationally. All of the trunk lines are FDM or TDM multiplexed. A local exchange switches all calls within its local network or transfers long distance calls to the nearest switching center. When a switching center receives a call, it either routes it to another local exchange within its area or passes it to another switching center. Note that a transmission path includes two local connections and may or may not utilize toll-connecting or intertoll trunks. A challenge for network designers is to keep signal amplitudes reasonably constant despite differences in path length.

The local connection, which is relatively short, is normally made with a pair of wires, twisted to reduce cross talk, called a twisted pair. A twisted pair may be used for distances up to approximately 8 Km.   Beyond this distance, loading coils are required to compensate for the capacitance between the wire pairs.   When the line attenuation exceeds 5 dB (at a distance of approximately 13 Km), an amplifier must be inserted into the line to restore the signal level to an acceptable value.   In general, the signal quality provided by a twisted pair is poor and such a system is only used for local connections due to its low cost.

Signal quality may be improved by using four wires rather than two.   One cable pair (amplified and loaded) is used for transmission and the other pair (also amplified and loaded) is used for reception.   Four-wire transmission systems are suitable for toll-connecting trunks between local exchanges and switching centers.

When a number of signals must be sent from one point to another, it is normally not economical to provide separate lines for each signal.   The most often used analog technique to minimize the number of lines required is frequency division multiplexing (FDM).   The FDM signal is inherently wide band and may only be sent over twisted-pair lines if loading coils are removed and amplifiers are inserted every 8 Km.   Since FDM requires expensive equipment, it is rarely economical for short links, unless installation of additional cables presents considerable difficulty.   Coaxial cable is sometimes used for long links because of its bandwidth capability, but line-of-sight microwave, troposcatter, satellite and optical fiber links are also used for portions of the intertoll trunks.

In addition to the frequency division multiplexed analog systems, the public telephone networks have developed digital carrier systems.   These systems use pulse code modulation (PCM) and time division multiplexing (TDM), rather than frequency division multiplexing.

19.11 PACKET SWITCHING SYSTEMS

A network utilized for data communications must provide a means of switching to provide a connection between the two points that wish to exchange data.   The existing voice switching facilities provided by the public telephone network allow for a switching technique known as circuit switching.   In a circuit-switched connection the lines (or TDM time slots) are dedicated for the entire communications session.   If high data rates prevail throughout the entire session, this technique is very efficient.   If the communication is low-speed or occurs in bursts, circuit switching may be quite inefficient.

Another switching technique that is advantageous in certain applications is message switching.   Message switching utilizes a buffer memory at the data switching exchange.   The message is transmitted when a transmission circuit becomes available.   This switching technique increases network efficiency, since lines can be time-multiplexed.   In addition, "broadcasting" to a number of different receivers is possible.

Packet switching is a more recently developed data communications technique, which divides data into small "bundles" or "packets" of information.   Each packet contains source and destination coding, the information bits, error control bits and network control information.   The packets may be held briefly at each data switching exchange, but are forwarded to their destination in near-real time. Packet networks are sometimes called "store and forward" networks.   Packet switching is most effective when users have low to moderate data rates and burst-like data communications requirements.   The data switching exchange controls the routing of individual packets over the network to maximize efficiency.   Two sequential packets may take vastly different routes from transmitter to receiver over the network, but since each packet has the necessary source and destination coding, this is transparent to the users.   Packet-switched networks are extremely efficient but require adherence to strict network protocols.

Protocols are internationally agreed-upon standards for data communications. The International Standards Organization (ISO) has defined a seven-layer Open Systems Interconnection (OSI) model.   This model may be used to define all aspects of data communications.   The seven layers are defined as follows:

Layer 1 (physical layer)—defines the electrical and mechanical interfaces between nodes in the network.

Layer 2 (data link layer)—ensures reliable transfer of data over the network and defines the procedures for error control.

Layer 3 (network layer)—determines the network configuration required for transfer of packets of data between nodes.

Layer 4 (transport layer)—provides message-transfer specifications between end users.

Layer 5 (session layer)—ensures the availability of appropriate resources for a session on the network.

Layer 6 (presentation layer)—ensures a common syntax for communication between different devices on the network.

Layer 7 (application layer)—responsible for general management of the network and interaction with human operators.

Examples of layer 1 protocols include standards, such as the RS-232C interconnection standard used widely in personal computers, as well as the higher performance RS-422/RS-449 standard.   These standards define electrical and/or mechanical interfaces.

The RS-232C interface is usually implemented with a 25-pin connector.   The connector pins are used for either control signals or digital data.   The driver voltage associated with a logical one may range from $+5$ to $+15$ volts, while a logical zero is represented by -5 to -15 volts.   At the receiver, $+3$ volts or more is interpreted as a logical one, and -3 volts or more negative is interpreted as a logical zero. The unbalanced or single-ended RS-232C interface suffers from some serious lim-

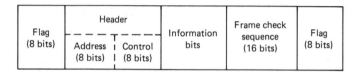

Figure 19-1 Frame structure for HDLC protocol.

itations, including the incompatibility with widely used 5-volt logic circuits, a maximum cable length of approximately 15 meters and a maximum data rate of 20 Kbps.

The RS-422 interface is balanced and provides significantly improved performance over the RS-232C interface. The mechanical specifications for the RS-422 interface are contained in the RS-499 specification. This specification calls for a 37-pin connector for primary channels and a 9-pin connector for secondary channels. An RS-422 interface may utilize cable lengths up to approximately 1.2 Km and is capable of a data rate of 10 Mbps for short cable lengths of 12 meters or less.

An example of a layer 2 protocol is the so-called High-Level Data Link Control (HDLC) as defined by the International Standards Organization. A subset of HDLC is IBM's SDLC, or Synchronous Data Link Control standard. HDLC applies to synchronous data transmission, as opposed to the alternative, asynchronous transmission. The primary difference between the two types of transmission is timing. Asynchronous transmission occurs on a character by character basis with start and stop bits for each character. The time between characters is variable. Synchronous transmission, on the other hand, occurs continuously with no start-stop bits once the transmitter and receiver are synchronized. HDLC is a bit-oriented protocol, as opposed to the character-oriented protocols used in asynchronous transmission. All bit-oriented protocols have a similar "frame" structure. Figure 19-1 illustrates the frame structure for an HDLC protocol.

The standard that defines the user to packet network interface was developed by the CCITT (Comité Consultatif International Téléphonique et Télégraphie) and is known as X.25. This standard addresses only layers 1, 2 and 3 of the ISO protocol hierarchy. X.25 uses existing standards, such as RS-232C and RS-422 for the physical layer and a version of HDLC for the data link layer. The network layer of X.25 specifies three types of switching services offered in the packet mode. These services are called the Permanent Virtual Circuit (similar to a slow leased line), Virtual Call (similar to a dial-up circuit), and the Datagram (used for single packet messages).

19.12 INTEGRATED SERVICES DIGITAL NETWORK

Integrated Services Digital Network (ISDN) describes a network concept in which voice, data, audio and video services, meter reading, alarms, control signals for energy management, and various other services are delivered over a single network.

ISDN is currently in a state of evolution in many countries and is based upon the Integrated Digital Network (IDN) currently evolving for the public telephone networks. IDN is based upon a PCM system with a sampling rate of 8 KHz, 8-bit encoding/decoding with either A-law or μ-law companding for a data rate of 64 Kbs. The ISDN will grow from the IDN by adding additional functions and features. ISDN will conform to the ISO seven-layer OSI model discussed in Section 19.11. Development of ISDNs is coordinated by the CCITT, but guidelines recognize that countries will develop different ISDNs, so emphasis is placed upon establishing compatible international interconnection standards.

In North America, because of the many commercial companies providing telecommunications services, no single unified plan exists for the development of an ISDN. Because of the competition, it is likely that multiple integrated networks will be developed. In any case, because of business constraints, it is unlikely that significant progress will be made toward a nationwide ISDN before the year 2000.

19.13 LOCAL AREA NETWORKS

The proliferation of personal computers in business, industry, and college campuses has resulted in pressure to provide communications capability for these machines. The result has been the development of a number of different types of local area networks (LANs) by various manufacturers. The Institute of Electrical and Electronic Engineers (IEEE) has developed a series of local area network standards (IEEE-802) to ensure compatibility between equipment made by different manufacturers. These standards address layers 1 and 2 of the ISO OSI model. Figure 19-2 illustrates the three most popular network topologies.

The star configuration is used when a number of terminals or computers need access to a dominant central node. The dominant central node may have characteristics such as high-capacity mass storage and/or high-speed processing. The disadvantage of the star configuration is the failure of the entire network if the central node fails.

The ring configuration avoids the problem of central node failure bringing down the entire network but introduces a different failure mechanism. If any one node in the ring fails, the network will go down, since information must be passed from node to node. Advanced implementations of the ring network provide for redundant paths and/or fail-safe bypassing of a defective node.

The bus configuration allows the failure of a node, provided the failure results in a high-impedance load on the bus. If the failure results in a short circuit, the entire network will become inoperative.

IEEE-802 specifies two types of bus-oriented networks and one type of ring-oriented network. The differences between the networks are a result of how the problem of contention (accessing the network) is handled. Two approaches to contention are addressed in IEEE-802: CSMA/CD (carrier-sense, multiple access

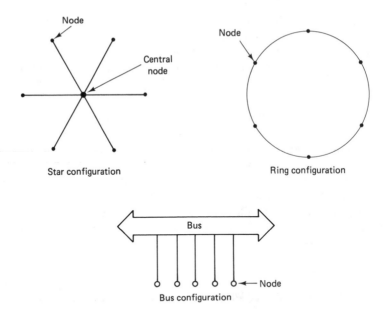

Figure 19-2   Local area network topologies.

with collision detection), and token-passing. The bus-oriented LANs utilize CSMA/CD and token-passing, while the ring network uses only token-passing.

In a LAN utilizing CSMA/CD, all nodes monitor the network and in the absence of any activity on the bus, any node may start to transmit. If two or more nodes transmit simultaneously, a collision occurs and retransmission after a random delay time is required. Nodes must monitor the network after beginning transmission to detect collisions. If the network is heavily loaded, multiple collisions will occur and the throughput will decrease. CSMA/CD systems are generally avoided where guaranteed access to the network is required.

The second approach to LANs is token-passing. A "token" is a digital word that is moved around the network from node to node. When a station (node) receives the token, it is able to transmit. After transmission, the token is passed to the next station. Token-passing ensures each station access to the network.

LANs may utilize either baseband or broadband transmission. Baseband transmission systems are single-channel systems and are less complex than carrier-based broadband systems, which utilize frequency division multiplexing to achieve multiple-channel capability. LANs are all basically packet networks in that they provide communications via short bursts of information organized in packets.

An example of an industry-standard local area network is Ethernet, which was developed by Xerox Corporation in conjunction with Digital Equipment Corporation and Intel Corporation. Ethernet is a CSMA/CD baseband system.

EXERCISES ━━━━━━━━━━━━━━━━━━━━━━━━━━━━━━━━━━━━

1. Obtain a technical manual for a communications system discussed in this chapter. Write a short report discussing all aspects of the selected system, including theory of operation, block diagram, and applications.
2. Obtain a recent trade journal related to communications. Write a short review of the journal evaluating the advertising, as well as the technical articles.

Bibliography

TECHNOLOGY LEVEL COMMUNICATIONS TEXTS

ALISOUSKAS, VINCENT F., and TOMASI, WAYNE. *Digital and Data Communications.* Englewood Cliffs: Prentice-Hall, 1985.

DEFRANCE, J. J. *Communications Electronics Circuits*, 2nd ed. New York: Holt, Rinehart and Winston, 1972.

DOUGLAS, ROBERT L. *Satellite Communications Technology.* Englewood Cliffs: Prentice-Hall, 1988.

KENNEDY, GEORGE. *Electronic Communication Systems*, 3rd ed. New York: McGraw-Hill, 1985.

KILLEN, HAROLD B. *Digital Communications with Fiber Optics and Satellite Applications.* Englewood Cliffs: Prentice-Hall, 1988.

KILLEN, HAROLD B. *Telecommunications and Data Communication System Design with Troubleshooting.* Englewood Cliffs: Prentice-Hall, 1986.

LENK, JOHN D. *Handbook of Data Communications.* Englewood Cliffs: Prentice-Hall, 1984.

MILLER, GARY M. *Modern Electronic Communication*, 3rd ed. Englewood Cliffs: Prentice-Hall, 1988.

SINNEMA, WILLIAM. *Electronic Transmission Technology: Lines, Waves, and Antennas.* Englewood Cliffs: Prentice-Hall, 1988.

TOMASI, WAYNE F. *Electronic Communications Systems: Fundamentals Through Advanced.* Englewood Cliffs: Prentice-Hall, 1988.

TOMASI, WAYNE F. *Fundamentals of Electronic Communications Systems.* Englewood Cliffs: Prentice-Hall, 1988.

TOMASI, WAYNE, and ALISOUSKAS, VINCENT F. *Telecommunications: Voice/Data with Fiber Optic Applications.* Englewood Cliffs: Prentice-Hall, 1988.

WICKERSHAM, A. F. *Microwave and Fiber Optics Communications.* Englewood Cliffs: Prentice-Hall, 1988.

ENGINEERING LEVEL COMMUNICATIONS TEXTS ▬▬▬▬▬▬▬▬▬

BEST, ROLAND E. *Phase-locked Loops Theory, Design and Applications*. New York: McGraw-Hill, 1984.

CARSON, RALPH. *High Frequency Amplifiers*. New York: Wiley, 1982.

CHORAFAS, DIMITRIS N. *Telephony Today and Tomorrow*. Englewood Cliffs: Prentice-Hall, 1984.

CLARKE, KENNETH K., and HESS, DONALD T. *Communication Circuits: Analysis and Design*. Reading, MA: Addison-Wesley, 1971.

DIXON, ROBERT C. *Spread Spectrum Systems*, 2nd ed. New York: Wiley Interscience, 1983.

EDWARDS, T. *Foundations for Microstrip Circuit Design*. New York: Wiley, 1981.

FEHER, KAMILO. *Digital Communications Microwave Applications*. Englewood Cliffs: Prentice-Hall, 1981.

FEHER, KAMILO. *Digital Communications Satellite/Earth Station Engineering*. Englewood Cliffs: Prentice-Hall, 1983.

FUSCO, VINCENT. *Microwave Circuits—Analysis and Computer-aided Design*. Englewood Cliffs: Prentice-Hall, 1987.

GONZALES, E. *Microwave Transistor Amplifiers—Analysis and Design*. Englewood Cliffs: Prentice-Hall, 1984.

HA, T. *Solid State Microwave Amplifier Design*. New York: Wiley Interscience, 1981.

LIAO, S. *Microwave Circuit Analysis and Amplifier Design*. Englewood Cliffs: Prentice-Hall, 1987.

LIAO, S. *Microwave Devices and Circuits*. Englewood Cliffs: Prentice-Hall, 1985.

MILLIGAN, T. *Modern Antenna Design*. New York: McGraw-Hill, 1985.

OWEN, FRANK F. E. *PCM and Digital Transmission Systems*. New York: McGraw-Hill, 1982.

PANTER, PHILIP F. *Communication Systems Design: Line-of-sight and Tropo-scatter Systems*. New York: McGraw-Hill, 1972.

PROAKIS, JOHN G. *Digital Communications*. New York: McGraw-Hill, 1983.

RODEN, MARTIN S. *Analog and Digital Communication Systems*, 2nd ed. Englewood Cliffs: Prentice-Hall, 1985.

ROHDE, ULRICH L. *Digital PLL Frequency Synthesizers Theory and Design*. Englewood Cliffs: Prentice-Hall, 1983.

SCHWARTZ, MISCHA. *Information Transmission, Modulation, and Noise*, 2nd ed. New York: McGraw-Hill, 1970.

TAUB, HERBERT, and SCHILLING, DONALD L. *Principles of Communication Systems*. New York: McGraw-Hill, 1971.

ZIEMER, R. E., and TRANTER, W. H. *Principles of Communications Systems, Modulation and Noise*. Boston: Houghton Mifflin, 1976.

MANUFACTURERS' DATA BOOKS ───────────────────────────

Databook RF and Power Semiconductors. TRW Semiconductor Division, Lawndale, CA, 1983.

Integrated Circuit Databook. Plessey Solid State, Irvine, CA, 1983.

Linear Databook. National Semiconductor Corporation, Santa Clara, CA, 1982.

Linear LSI Data and Applications Manual. Signetics Corporation, Sunnyvale, CA, 1985.

Linear Supplement Databook. National Semiconductor Corporation, Santa Clara, CA, 1984.

RF Device Data. Motorola Inc., Phoenix, AZ, 1983.

Standard Product Guide. EXAR Corporation, Sunnyvale, CA, 1984.

Index

Frequency translator, 53
Fresnel clearance, 282
Friis transmission formula, 278
Fringe field, microstrip antenna, 271

GaAs MESFET detectors, 297
Gain, antenna, 253
Gallium arsenide field effect transistors
    (GaAs FET), 223
Graded index cable, 294
Gray code, 68
Guard band, 32, 53

Half-cosine waveform, 12–13
Hamming codes, 82
Hartley-Shannon law, 17
Helix antenna, 262
High pass filter, 33
Horizontal polarization, 254
Horn antenna, 265
Hot carrier diode, 180
h-parameters, 201
H-plane, 251
Hybrid, quadrature, 208
Hyperbolic tangent function, 190

Impedance transformation, 149
Impedance, characteristic, 187
Index of refraction
    atmosphere, 286
    optical cables, 290
Infinite gain multiple feedback filter, 134
Information carrying capacity, 17
In-phase channel, 99
Input impedance, transmission line, 190
Insertion loss, 33
Integrated services digital network
    (ISDN), 308
Intermediate frequency amplifiers (IF
    amplifiers), 229
Intermediate frequency (IF), 213
Inverse Chebyshev filter, 125
Ionospheric propagation, 287
Isolated port, 207
Isolation, 219
Isolator, 210
Isotropic antenna, 251

Jamming margin, 114
Justification, 71

Kelvins, 20

Laser diode sources, 296
Light emitting diode sources (LED
    sources), 296
Limiter circuits, 229
Linear phase locked loop, 160
Line-of-sight, 278
Line-of-sight systems, 303
Load reflection coefficient, 224
Local area networks (LAN), 309
Local oscillator, 219
Lock range, 161
Longitudinal redundancy check (LRC),
    80
Loop filter, 160
Loss factor, 26
Loss
    conductor, 242
    dielectric, 242
Lower sideband (LSB), 38
Low pass filter, 33
Low-noise amplifiers, 223

μ-law, 65
M-ary phase shift keying modulation, 103
Matching networks, 151
Material dispersion, 293
Maximal linear codes, 120
Maximum power transfer, 20
Maximum usable frequency (MUF), 288
Maxwell's equations, 251
Microstrip, 191
    antenna, 270
    filter, 210
    transmission lines, 248
Minimum shift keying modulation
    (MSK), 107
Mismatch, 187
Mixer circuits, 219
Mixers, active, 219
Mixing, 37
Mobile communications systems, 301
Modal dispersion, 293